Energy Futures

Energy Futures

Edited by
Ralf Boscheck

palgrave
macmillan

First published 2007 by
PALGRAVE MACMILLAN
Houndmills, Basingstoke, Hampshire RG21 6XS and
175 Fifth Avenue, New York, N.Y. 10010
Companies and representatives throughout the world

PALGRAVE MACMILLAN is the global academic imprint of the Palgrave Macmillan division of St. Martin's Press, LLC and of Palgrave Macmillan Ltd. Macmillan® is a registered trademark in the United States, United Kingdom and other countries. Palgrave is a registered trademark in the European Union and other countries.

ISBN-13: 978-0-230-01367-4 hardback
ISBN-10: 0-230-01367-8 hardback

This book is printed on paper suitable for recycling and made from fully managed and sustained forest sources.

A catalogue record for this book is available from the British Library.

Library of Congress Cataloging-in-Publication Data
Energy futures / edited by Ralf Boscheck.
p. cm.
Papers of a conference held by the IMD in Nov. 2005.
Includes bibliographical references and index.
ISBN-13: 978-0-230-01367-4 (cloth)
ISBN-10: 0-230-01367-8 (cloth)
1. Energy industries—Economic aspects—Congresses. 2. Power resources—Economic aspects—Congresses. 3. Energy policy—Economic aspects—Congresses. I. Boscheck, Ralf, 1959–

HD9502.A2E543934 2007
333.79—dc22

2006047189

10 9 8 7 6 5 4 3 2
16 15 14 13 12 11 10 09 08 07

Printed and bound in Great Britain by
Antony Rowe Ltd, Chippenham and Eastbourne

The book is dedicated to the memory of

Adolf H. Lundin

a true oilman whose analytical skills, self-discipline and courage to take risks made him an outstanding entrepreneur and a genuine visionary of political turns and surprising market developments.

Adolf H. Lundin is the founder of the Lundin Group which consists of several public companies engaged in exploration for and production of oil, gas and metals on a global scale. He was nominated Swede of the Year in 1998 and became an Honorary Doctor of the Plekhanov Russian Academy of Economics in 2002.

Contents

Part Three Downstream

Part Four Beyond the Mere Technical – A Broader Perspective

List of Figures and Tables

Figures

Tables

List of Common Abbreviations

/bbl	Per barrel of oil
boe	Barrels of oil equivalent
CAPEX	Capital expenses
/d	Per day
E&P	Exploration and production
ERR	Effective royalty rate
IEA	International Energy Agency
IOC	International oil company
kWh	Kilowatt-hour
LNG	Liquefied natural gas
MM	Million
MMbpd	Million barrels per day
MWh	Megawatt-hour
NOC	National oil company
OPEX	Operating expenses
PSA	Production sharing agreement
PSC	Production sharing contract
R&D	Research and development
R/T	Royalty/tax system
ROCE	Return on Capital Employed
ROR	Rate of return
SA	Service Agreement
toe	Tons of oil equivalent
TWh	Terawatt-hour

Acknowledgments

Energy Futures is the result of a team effort that led 18 international energy researchers, policymakers and executives to gather at the International Institute of Management Development (IMD) in November 2005. The meeting had been inspired by Adolf and Ian Lundin who endowed the Lundin Family chair at the institute to sponsor energy-related research in business policy and economics. This publication brings together an eclectic set of analyses and contributions as a first sketch of the territory. A range of specific research initiatives have since emerged whose outcomes will be published in different venues.

In preparing, conducting and following up on the event three colleagues at IMD were invaluable. Working from an original invitation list, Mope Ogunsulire undertook background research, contacted contributors and added the section with reference material at the end of this book. An engineer by training and a World Banker by vocation, she brought analytical skills, political awareness and a good sense of humor to the project. Anne-Séverine Vinet organized the logistics, administered the overall program and got the "cast" safely to and back from Lausanne. Virginie Boillat-Carrard worked hand-in-hand with Palgrave Macmillan to see the manuscript through to completion. Outside of IMD, Caroline Taggart reviewed all papers and provided valuable editorial comments. I would like to thank all of them.

Lausanne, June 2006

Notes on Contributors

Ralf Boscheck, Lead author and editor
Lundin Family Professor of Business Policy and Economics

PhD Economics, Universität St. Gallen, Switzerland; MBA Katholieke Universiteit, Leuven, Belgium; MA Georgetown University, Washington D.C. His research interests are industry and market analysis with a particular focus on energy, healthcare and media. His research on regulation, competition and economic policy has been published in *World Competition*, *Intereconomics*, *California Management Review* and *Long Range Planning* as well as in books and the international business press. Before joining IMD in 1991, Ralf Boscheck was a senior consultant with Monitor Company to Fortune 500 companies across a range of sectors assisting in the formulation and implementation of business units and corporate strategies. He has advised national and international organizations and governments in the areas of market liberalization, industrial strategy and competition policy.

Christine Batruch
Vice President Corporate Responsibility, Lundin Petroleum

Hon. BA, BCL, LL.B., barrister and solicitor (University of Toronto, McGill University, Law Society of Upper Canada). Based in Geneva, Christine has worked in the non-profit, academic and business sectors. From 1992 to 1995, she participated in the establishment of a number of non-profit institutions in the Ukraine part of the Soros network of foundations. From 1995 to 1999, she worked at the International Academy of the Environment as executive assistant to the director and later assistant professor. She joined Lundin Oil in 1999 and became Vice President Corporate Responsibility at Lundin Petroleum in 2002.

Subhes C. Bhattacharyya
Senior Lecturer, CEPMLP, University of Dundee, Scotland

PhD Applied Economics (specialization in Energy Economics), DEA Applied Economics, University of Grenoble, France; ME Energy Planning and Policy, Asian Institute of Technology, Thailand; BE Mechanical Engineering, Bengal Engineering College, University of Calcutta. Dr.

Bhattacharyya specializes in energy economics, energy planning and policy issues, regulation and restructuring of energy industries and energy-environment interactions. He has a combination of graduate-level teaching, research and industry experience of more than 14 years. At present, he is a senior lecturer at the Centre for Energy, Petroleum, and Mineral Law and Policy (CEPMLP) at the University of Dundee. Earlier, he was a visiting associate professor in Energy Program of the Asian Institute of Technology and an associate professor at the Madras School of Economics in India. He has worked as a consultant with PricewaterhouseCoopers, and as an economist with the National Economic Research Associates. He has published many papers in leading energy journals. He has also worked as one of the principal investigators in a number of research projects funded by government, firms and international funding agencies.

Stefan Bockamp
Coordinator Emissions Trading, E.ON Energie AG

PhD Mechanical Engineering, Siegen University, Germany. Since 2005, Dr.-Ing. Bockamp has been E.ON Energie's coordinator for emissions trading, responsible for coordination of E.ON Energie group-wide emissions trading activities. Prior to that, he was a project manager for R&D projects for new power plant concepts and the implementation of new regulations in power plants. He has published numerous conference proceedings and articles.

Sergio Castedo
Business Development Manager, Power Europe and Asia with BP Alternative Energy

Executive MBA, European School of Management ESCP-EAP, France; MSc Energy Management, Norwegian School of Management; BSc Mechanical Engineering, University of Maryland, USA. Mr. Castedo is currently Business Development Manager for grid-power in Europe and Asia with BP Alternative Energy. At the time this paper was written Mr. Castedo was project manager in the International Development Department of Statkraft, Norway's largest electricity producer and Europe's second largest producer of renewable energy. At Statkraft, he was responsible for business development in renewables. Prior to that, he worked for five years with GE Wind (formerly Enron Wind) in the United States, Latin America and Spain where he closed over $200 million in wind power contracts. Before moving to Enron Wind, he worked for the American Wind Energy

Association in Washington D.C. During that time, he also worked with the US National Renewable Energy Laboratory in Colorado.

David R. Demers
CEO, Westport Innovations Inc.

BSc Physics, BSc Law, University of Saskatchewan, Canada. David Demers is founder and chief executive officer of Westport Innovations. Based in Vancouver, British Columbia, Canada, Westport Innovations is dedicated to leading the transition of the international commercial engine industry from oil-based to gaseous fuels. The company develops technologies that allow diesel engines to operate cleanly and efficiently on alternative fuels such as natural gas, propane and hydrogen. Mr. Demers has been president since the company was formed in March 1995. Before that, he worked for IBM Canada and then founded and served as president of a closely held consulting company specializing in software marketing, finance and business transformation. Mr. Demers is also a member of the board of directors of Cummins Westport Inc. and Clean Energy Inc., private companies in which Westport has an investment interest.

Carlos A. Garibaldi
Managing Director, Harrison Lovegrove & Co.

MBA, Rice University (Jones Scholar, certificate of specialization in M&A negotiations), USA; MSc Petroleum Engineering, University of Tulsa, USA; MSc, BSc Chemical Engineering, Universidad Nacional de Buenos Aires, Argentina. Carlos A. Garibaldi is a managing director of Harrison Lovegrove & Co. in charge of the Houston office. Until October 2005, Mr. Garibaldi was a senior vice president and board director in The Scotia Group, Inc., based in Houston. He has over 26 years of industry experience in reservoir engineering and research, operations, economic modeling, capital budgeting, strategic planning, business development, contractual and fiscal terms, negotiations, and management consulting. His interest is focused mainly on Latin America. Before joining the Scotia Group, Mr. Garibaldi was with Amoco (reservoir, research and operations engineer, reserves and budget planner), Plains Resources (engineering and acquisitions manager, vice president and general manager for International), Arthur D. Little (senior manager), San Jorge International (president and COO), Chevron Latin America (vice president of business development), and G&G Energy Consultants (partner). He teaches courses on petroleum economics, strategic planning, contractual and fiscal terms and negotiations in six Latin American countries.

Gunnar Gjerde
Director General, Oil and Gas Department, Norwegian Ministry of Petroleum and Energy

Law degree from the University of Oslo, Norway. Mr. Gjerde has held various positions in the Ministry of Petroleum and Energy since 1973, and has wide experience from the oil and gas department's field of action ranging from the award of production licences, development policies for both oil and gas, legal questions, oil market questions and industrial and technology issues. From 1992 to 1995, Mr. Gjerde was stationed in Brussels working in the Norwegian Delegation to the European Union. In the 1970s, he held positions in the Ministry of Environmental Affairs and the Norwegian Agency for Development Cooperation (NORAD). He has also held a position as a deputy judge. He was appointed head of the oil and gas department in May 2001, and became the director general of the exploration and development market department in August 2001. From July 2005, he has been the director general of the oil and gas department.

Karen Hall
Vice President, Technical Operations, National Hydrogen Association

MSc Environmental Science, Johns Hopkins University, USA; BSc Welding Engineering, Ohio State University, USA. Karen Hall is the managing director for Technology Transition Corporation, Ltd. in Gateshead, UK. She performs management and technical services for organizations interested in commercializing alternative energy systems. Ms. Hall serves as vice president of technical operations for the National Hydrogen Association (NHA), USA. She plans and conducts workshops globally to educate stakeholders on the market and technical aspects of hydrogen energy and fuel cells. Since 1997, Ms. Hall has organized Hydrogen Safety, Codes and Standards Workshops for the NHA. These workshops develop consensus among the stakeholders on technical issues pertinent to the commercial deployment of hydrogen energy systems. They also serve to create dialog between hydrogen industry, code and standards development organizations, code officials, and other stakeholders for the purpose of achieving common objectives. Ms. Hall participates in the development and management of industry-led associations, particularly in the areas of hydrogen and fuel cells. She is familiar with the stakeholders and the processes used to create consensus among diverse stakeholders.

Geir Holler
Senior Vice President for Energy Management, Statkraft AS

MSc Economics; BSc Economics (majoring in international economics and trade theory), University of Oslo, Norway. Mr. Holler is a senior vice president for energy management at Statkraft. He has 11 years' experience in Statkraft, and has held line management positions for different parts of Statkraft's market operations including market analysis, energy trading and risk management, and energy management. Prior to joining Statkraft in 1995, he gained nine years' experience with Norconsult International as an international consultant, project manager, and energy and transport economist on various energy and industry projects in Europe, Asia, Africa and the USA.

Bjørn N. Holsen
Vice President, International Development, Statkraft AS

MSc Industrial Engineering, Technical University, Trondheim, Norway. Bjørn N. Holsen joined the project development division of Statoil in 1993 and held various positions as project controller. In 1997, he joined Naturkraft AS, a Norwegian CCGT project developer jointly owned by Statoil, Norsk Hydro and Statkraft, first as project control manager, and later as commercial director. In 2001, he joined Statkraft and worked for two years in its Amsterdam based trading office as financial manager. He is currently responsible for launching Statkraft's international project development activities, with an initial focus on wind power development in the UK.

David Johnston
Director, Daniel Johnston & Co. Inc. USA

BSc Electrical Engineering, University of Rochester, New York. David Johnston is an engineering and petroleum industry consultant who has worked and lectured internationally on the subjects of energy and technology. He has published numerous articles, and two books, *Maximum Efficient Production Rate* and *Economic Modelling and Risk Analysis Handbook* with the University of Dundee, Scotland, and "Introduction to Oil Company Financial Analysis" with PennWell. Mr. Johnston is a director with Daniel Johnston & Co. Inc. and works with oil companies, national oil companies, and governments in the design and analysis of petroleum fiscal systems.

Jennifer Josefson
Associate, White & Case LLC, Russia

BA Russian Language and Foreign Service, Baylor University, USA; JD Law, University of Texas School of Law, USA; Masters degree in Law, Institute of State and Law, Russian Academy of Sciences. Jennifer Josefson is a lawyer in the Energy Practice Group of White & Case's Moscow office. White & Case is a global law firm, with nearly 2,000 lawyers in 25 countries around the world. White & Case in Moscow is one of Russia's leading international law firms, and was named "Russian Law Firm of the Year" by the International Financial Law Review in 2005. Ms. Josefson represents international and Russian oil companies, oilfield service companies and other corporate clients active in Russia's oil and gas sector. She has advised on a number of international transactions and production sharing agreements, and has negotiated mergers and acquisitions, joint ventures, joint-operating agreements, supply contracts and various oilfield service agreements. Before joining White & Case, she worked for a large US law firm in Houston, Texas. Her articles on Russian oil and gas issues have been published in several journals and magazines. Ms. Josefson is a member of the American Bar Association and was vice-chair of the Eurasian States Committee of the International Law Section of the ABA from 2002 to 2003. She has been a member of the Association of International Petroleum Negotiators' (AIPN) board of directors since 2004 and is also a member of the EAMC steering committee, coordinating AIPN's activities in Moscow and a co-chair for AIPN's 2005 international conference. Ms. Josefson is admitted to the State Bar of Texas.

Jörg Kruhl
Head of Technology Policy, New Technologies, E.ON Energie AG

PhD Mechanical Engineering, University of Hannover, Germany. His thesis was entitled "Application of Fuel Cells for Domestic Heat and Power Supply." Dr.-Ing. Kruhl joined E.ON in 2000 and held a series of managerial positions as asset and project manager, responsible for long-term strategic and capacity planning. He led the E.ON Energie group-wide implementation of emissions trading and then became head of the department "Technology Policies and New Technologies" in May 2006.

Olav Ljosne
Regional External Affairs Manager, Shell Exploration and Production

Degrees in Political Science, Sociology, Law and Arabic, University of Oslo, Norway. Olav Ljosne has held executive and elected positions in national Norwegian (Liberal Party) and European politics. Also, he has worked for the Norwegian government in Trade and Bilateral Commercial Relations and UN relations, and held diplomatic positions at the Norwegian embassies in Iraq, Saudi Arabia and Iran. In 1990, he joined Shell, working in Norske Shell Public Affairs, Shell International Group Public Affairs – Middle East and Africa, Svenska Shell, Shell Europe Oil Products (Nordic Area, Russia and Baltics) and until 2005 he was with Shell International Exploration and Production. In 2005, he became the regional external affairs manager for Shell Exploration and Production Africa.

Caryl Pfeiffer
Director, Corporate Fuels and By-Products LG&E Energy

BA (cum laude) Biology/Business Administration, Wittenberg University, USA; Graduate Fellowship Award, Department of Landscape Architecture, University of Illinois, USA; Graduate work, School of Natural Resources (emphasis on economics and energy development), Ohio State University, USA. Caryl Pfeiffer is director of corporate fuels and by-products for LG&E Energy (E.ON USA) responsible for developing and implementing strategies for the use of solid fuels within the integrated energy business. Prior to that, she was director for LG&E Energy's environmental affairs from 1998 to 2003. Before joining LG&E Energy in 1998, she held a series of positions in environmental affairs with the Kentucky Utilities Company and the Columbus and Southern Ohio Electric Company. She has over 26 years' experience in energy and environmental affairs and sits on various state and federal level environmental and energy committees, including the American Gas Association's Environmental Regulatory Action Committee, the Edison Electric Institute Environment and Energy Committee, and the US Environmental Protection Agency's FACA Subcommittee on Ozone, Particulate Matter, and Regional Haze.

Álvaro F. Racero
Director North America, Repsol YPF

BSc Mining Engineering (specialization Geology and Geophysics), Royal School of Mines, Madrid, Spain; PhD (Honors), Polytechnic University of Madrid. His thesis was entitled "Structural and Sedimentary Evolution of the Intracratonic Tertiary Basins in Central Spain." Álvaro Racero started his professional career in the oil industry in 1985 as operations geophysicist with Prakla-Seismos. In 1986 he joined Shell International E&P, where he worked in Europe as exploration and production geologist and seismologist in the Mediterranean, Cantabrian and North Sea Basins. In 1994, he was posted to Venezuela to lead the evaluation team in the Urdaneta West asset, in the prolific Lake Maracaibo Basin, one of the first blocks of the Venezuela "apertura" awarded to Shell Venezuela S.A. Between 1997 and 2000, he was involved in several Shell projects in northern South America and the Caribbean, as the head of new ventures, based in Caracas. In early-2000, he became the head of exploration planning and portfolio management for the Brunei Shell Companies in Borneo, in the Far East; being an active contributor to some of the first deepwater wells drilled in the Baram Delta. Mr. Racero joined Repsol YPF in Madrid at the end of 2001 as the coordinator for corporate new ventures, with worldwide responsibilities. In April 2003, he was appointed director of exploration and development for the Caribbean Region, based in Houston, which encompassed Repsol YPF operations in the Gulf of Mexico, Mexico, Trinidad and Tobago, Venezuela and Guyana and Suriname. With the growth in activities in the US sector of the Gulf of Mexico and onshore Mexico, he was appointed director of the newly-created North American business unit in January 2004.

Aruna Subramanian
Senior Investment Analyst, Shell Technology Ventures

BTech, Civil Engineering, Indian Institute of Technology, New Delhi, India; MBA, IMD, Lausanne, Switzerland. Aruna Subramanian is a senior investment analyst with Shell Technology Ventures. Before joining Shell in 2005, Ms. Subramanian had over 11 years of technical, human resources and general management experience with Schlumberger Oilfield Services. Schlumberger is the leading multinational technology and information solutions provider in the energy sector.

Part One

Context

1
Energy Futures – Introduction and Overview

Ralf Boscheck

Despite prices of around $60 a barrel, the world is experiencing the fastest growth in oil demand in more than 25 years. Windfall profits rise, as do concerns about macro and microeconomic implications. What will be the effects on growth, investments and inflation in net-importer and producer countries? What will be the impact on sluggish exploration activities, reserve replacements, capacity expansions and hence energy supply? Will national resource owners adjust their development strategies? Are entirely new business models required? Will markets and regulatory initiatives deliver effective fuel substitution? To address some of these issues, the International Institute for Management Development (IMD) organized a research conference entitled "Energy Futures" in November 2005. Eighteen international energy researchers, policymakers and executives discussed specific aspects of today's energy industry and its likely evolution. This introduction briefly sketches various factors shaping our energy future in order to organize the conference contributions and express a common concern.

The price of oil

Any of a range of potential energy futures reflects the interaction of three main drivers: the development of oil and gas prices, opportunities for relaxing oil supply constraints and, finally, the capacity for speeding up fuel conservation and substitution to manage demand and greenhouse gas (GHG) emissions. Current high crude oil prices, in real terms, are still well below oil price peaks of the 1970s. But how stable are these prices? Oil price assumptions affect consumption and production and hence the price of oil. The fact that over the last two years, a variety of highly regarded analysts have often had to revise their forecasts has given rise to the notion that prices are purely speculative. But prices are forever speculative – what is important is what drives speculation.

In its most recent reports,[1] the International Energy Agency (IEA) predicts a 51% growth of energy consumption over the next two decades up to the year 2030. Oil is expected to remain the energy source of choice – its demand will continue to grow at around 1.9% annually. This development is fueled by sustained growth in Asia's emerging economies and the industrialized world as well as the global increase in transportation. To check the agency's predictions, one merely needs to derive energy demand from fairly accepted statistics regarding population growth, and anticipated levels of economic activity and energy intensity. Assuming that the world population increases by 1.4% annually, the global economy expands by 3.5% and technological progress reduces energy intensity by 2% on average, global energy demand will rise by 1.6% annually. Hence, by 2050 the world will require double its current level of energy. These numbers roughly square with IEA forecasts.

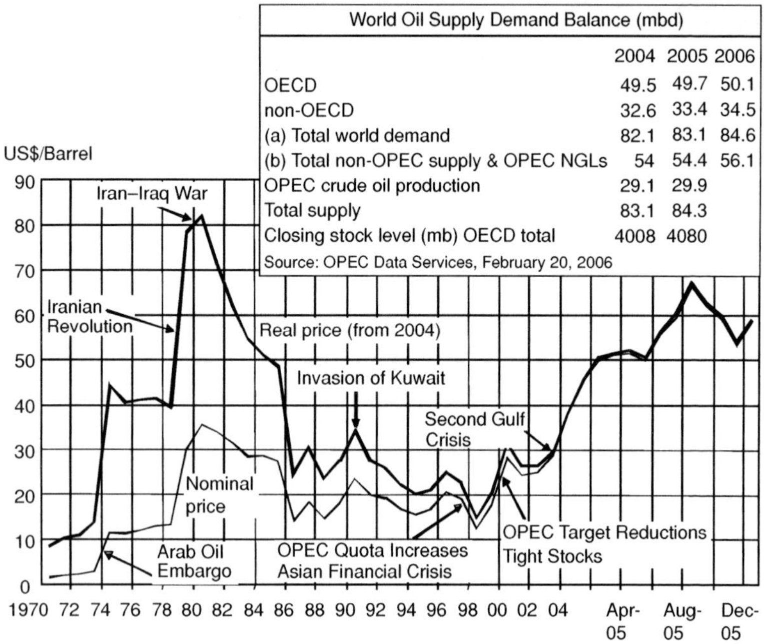

World Oil Supply Demand Balance (mbd)

	2004	2005	2006
OECD	49.5	49.7	50.1
non-OECD	32.6	33.4	34.5
(a) Total world demand	82.1	83.1	84.6
(b) Total non-OPEC supply & OPEC NGLs	54	54.4	56.1
OPEC crude oil production	29.1	29.9	
Total supply	83.1	84.3	
Closing stock level (mb) OECD total	4008	4080	

Source: OPEC Data Services, February 20, 2006

Source: Adapted from BP, Datastream, OPEC

Figure 1.1 Oil Price Evolution

Against this surge in demand stands a massive, impending gap in oil supply. If all currently known reserve capacities are included – i.e. existing reserves are adequately developed, industry pushes for more enhanced oil recovery, so-called non-conventional oils are developed and new discoveries are made – the world will still be short of roughly 62 million barrels per day (+/– 25%) by 2020. Put differently, in 14 years *and* in the absence of any major demand and supply-side adjustments, the world will need to find the equivalent of one additional UK North Sea – annually. Do current prices already factor in the substantial risk of non-achievement, viz. non-correction or, conversely, are prices still too low to drive the required change? What conditions drive supply and demand adjustments?

Supply-side adjustments – from macro to micro

At the outset one may wonder why, given limited short-term opportunities for fuel substitution as well as increased oil and share prices, oil producers would be interested in expanding supply at all. Two reasons may be given. First, sustained high oil prices would speed up initiatives for fuel substitution. Second, producers – countries and corporations alike – may no longer, if they ever did, form a sufficiently homogeneous block that could easily coordinate a common supply position. Put differently, the main question may not be whether producers are *willing* but rather whether they are at all *able* to expand supply in the short or medium term, and, if so, *how*.

Figure 1.2 gives a quick overview of the geographic base, the cost and the coordination of oil supply. Roughly 30% of total oil production and two-thirds of the known global reserves[2] are located in the Middle East; Saudi Arabia's largest field not only holds a production cost advantage but the country's national oil company tops the list of the largest, and largely state-owned, resource owners. In 2005, OPEC's total crude oil production of 29.9 million barrels per day (MMbpd) represented 35.5% of total supply; in 2004, seven of the 14 countries that produced more than 2 MMbpd and ten of the 14 countries exporting more than 1 MMbpd of total oil net were OPEC members. At 2004 crude oil production rates, OPEC's overall reserve to production ratio translates into 83 years of remaining reserves; non-OPEC producers hold a stock lasting 26 years.

In mid 2005, OPEC was estimated to be able to produce an additional 1.5 MMbpd, roughly 5.4% of its members' quotas but not enough to satisfy the IEA-estimated annual demand growth of 2 MMbpd. At the same time, Saudi Arabia's national oil company, Saudi Aramco, calculated the

(a)

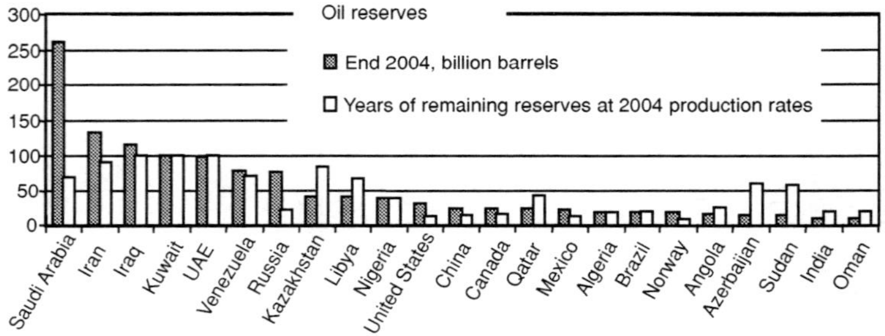

(b)

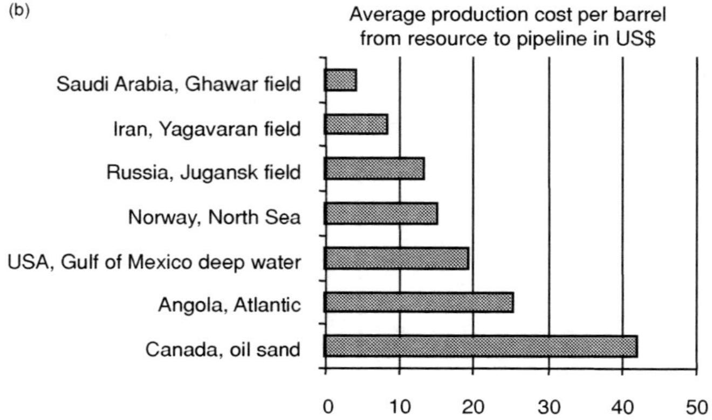

Source: Adapted from CERA and Petroleum Intelligence Weekly

(c) Top 20 oil companies, by reserves, 2003

		State Ownership %	million barrels
Saudi Aramco	Saudi Arabia	100	259,400
NIOC	Iran	100	125,800
INOC	Iraq	100	115,000
KPC	Kuwait	100	99,000
PDV	Venezuela	100	77,800
Adnoc	UAE	100	55,200
Libya NOC	Libya	100	22,700
NNPC	Nigeria	100	21,200
Pemex	Mexico	100	16,000
Lukoil	Russia	8	16,000
Gazprom	Russia	73	13,600
Exxon Mobil	USA	-	12,900
Yukos[a]	Russia	-	11,800
Petro China	China	90	11,000
Qatar Petroleum	Qatar	100	11,000
Sonatrach	Algeria	100	10,500
BP	Britain	-	10,100
Petrobras	Brazil	32	9,800
Chevron Texaco[b]	USA	-	8,600
Total	France	-	7,300

[a] Now in effect controlled by government

[b] Does not include acquired Uixal

Figure 1.2 Oil Reserves, Costs and Ownership

respective gap to be around 4.5 MMbpd. Of course, increased prices fuel investments to boost exploration and production and could conceivably lead to an excess-supply potential by the end of this decade. In fact, in mid 2005, Cambridge Energy Research Associates (CERA) predicted that global production capacity would increase to 101.5 MMbpd in 2010, i.e. resulting in a 7 MMbpd supply buffer on top of expected demand. But facts may be different and their implications not as clear cut.

Throughout the 1980s and 1990s low, volatile prices inhibited investments; today, OPEC member countries are set to raise crude capacity by about 5 MMbpd to 37.9 MMbpd in 2010. In the Gulf, the largest additions will come from Saudi Arabia, (+1.5 MMbpd), the UAE (+0.9 MMbpd) and Kuwait (+0.6 MMbpd). Libya is courting international oil companies (IOCs) to help raise its current capacity of 1.7 MMbpd back to the 3.3 MMbpd that it produced prior to international sanctions; Algeria is planning to add one-third to its current 2 MMbpd by 2010. While some of these expansions may strain OPEC quotas, countries like Venezuela and Indonesia are struggling to maintain even current output levels. Also, although increases in non-OPEC production, most importantly from countries like Angola, Brazil, Russia and the states bordering the Caspian Sea, are forthcoming, they are likely to have only a limited, short-term market impact. In fact, at the end of the decade, total non-OPEC oil production is projected to plateau and gradually decline, creating a production shortfall that upcoming Middle Eastern capacity expansion will barely meet.

In view of this, an excess supply potential by the end of this decade is highly unlikely. But even if it were to come about, prices need not come down. For one, limited construction and equipment capacities continue to drive up input costs – from steel prices to rig and tanker rates – and thereby require higher oil prices to justify investments. Next, once created, additional capacity merely provides an option to increase supply. Whether and how it would be used depends on the effectiveness of coordinating broadly different producer interests and views on potentially accelerating resource depletion across diverse macroeconomic and political conditions.

Given these demand and investment trends, any immediate price reduction would require the world's only swing producer, Saudi Arabia, to release its thinning spare capacity, cut profits and capital expenses (CAPEX). In the process, OPEC would lose a vital means of ensuring cartel cohesion and market management – an altogether highly improbable outcome. Adding to this the likelihood of further political tension or weather-based supply disruptions, one must expect prices to stay high.[3]

Even more disturbing, there is a risk of not meeting expansion targets at all. Changes in domestic agendas may undermine investments or further increase the output gap. What are the implications for resource owners, national oil companies and international operators?

Table 1.1 Hydrocarbon Exporters and Exposure

	Hydrocarbon net exporting countries' oil and natural gas exports and fiscal revenues in 2003		
	Oil and gas exports as a % of world oil and gas exports	Oil and gas exports as a % of country's GDP	Country's fiscal oil and gas revenues as share of country's GDP
Saudi Arabia	13.5	38.3	28.1
Russian Federation	11.8	17	6
Norway	6.5	18.4	12.2
United Arab Emirates	4.7	36.8	35.8
Iran, Islamic Rep.	4.3	19.8	16.3
Nigeria	4.3	46.1	28
Algeria	3.9	36.2	26.4
United Kingdom	3.8	1.3	0.4
Netherlands	3.4	4	n.a.
Venezuela, Rep. Bol.	3.3	24.6	23
Kuwait	3	44.8	48.1
Mexico	3	3	7.9
Indonesia	2.4	6.3	4.5
Canada	2.4	1.7	0.2
Libya	1.9	47.6	40.5
Iraq	1.6	38.4	n.a.
Qatar	1.5	47	24.9
Oman	1.5	43.1	35.4
Angola	1.4	65.3	28.3
Malaysia	1.2	7.4	4.3
Kazakhstan	1.1	23.6	6.2

Source: Adapted from IMF

Across all producers, price increases have resulted in increased earnings. According to the US Department of Energy, OPEC alone earned $300 billion in 2004 and $380 billion in 2005. Relative to the spending excesses of earlier oil-price bonanzas, most established producers appear to better control their use of windfall-profits. The IMF suggests that the average oil-exporting government is currently saving about two-thirds of the extra revenues.[4] Russia, for instance, repaid a $3.3 billion IMF loan in 2005 and followed the Norwegian example of transferring excess earnings into a stabilization fund to avoid overheating the economy. Similarly, Middle Eastern oil exporting countries – mindful of growing population and unemployment numbers – turned their fiscal deficits, averaging 3.5% of GDP in the mid 1990s, into a surplus of 7.9% of GDP at the end of 2004. And yet, there is reason for concern. Growing popular discontent with recent Russian welfare reforms or the political and economic consequences of maintaining high levels of income inequality while overheating Middle Eastern equity and real estate markets could bring prudent public finance to a halt and negatively affect the level of oil infrastructure investments. While such a perspective seems unwarranted with regard to some producer countries it is an unfortunate reality in others. Consider the case of Venezuela.

Holding 8% of the proven world oil reserves,[5] Venezuela is undergoing a constant transformation. Hugo Chavez's agenda to open the economy, counter corruption, make the state oil company more accountable and improve income distribution brought popular backing but still lacks the prerequisite institutional base. Hence, Venezuelan oil revenues fund political support and national welfare programs, while infrastructure investments by Petróleos de Venezuela are reportedly too low to boost needed production. Political rhetoric aside, the economy relies to a great extent on IOCs' investments. Their interest appears unbroken even after April 2005, when Chevron-Texaco, BP and Royal Dutch Shell were given six months to switch to government controlled joint venture contracts and accept an increasing income tax rate. But will investments be forthcoming in the event that operations are seriously politically constrained or US market access is blocked? Will markets cushion any perceived domestic squandering?

Venezuela's experience sheds light on how markets balance the interests of host countries and IOCs, and the roles that national oil companies (NOCs) may play. Resource owners typically seek investments, stable periodic income and a fair share of profits in addition to employment creation and technology transfer; IOCs look for resource access, low levels of transaction costs and government takes in line with prospects. Consequently, mature regions, like the North Sea, need to employ sophisticated fiscal regimes just to stay attractive and maintain exploration and production activity. In effect, it is largely due to governmental initiatives that the relatively low-prospect UK shelf attracted the second highest level of exploration activity in 2005. Decommissioning was delayed and the country can expect to be self-sufficient for 50% of its total oil and gas needs up to 2020. Conversely, much more promising regions cannot avoid creating a "feeding frenzy" among investors, even if, as in the case of Venezuela or Nigeria, political conditions are shaky and investments appear at risk. Of course different oil companies react differently.

The five global, integrated oil majors, Exxon-Mobil, BP, Shell, Chevron-Texaco and Total, accounting for 17% of the world market in 2004, operate alongside a group of non-integrated exploration and production companies – consisting of a few "super-independents" and a myriad of small ones. While both groups must be concerned with operating efficiency and balance exploration and acquisition in replacing and growing reserves, majors are also driven by scale-requirements to look for more and bigger projects and the need to protect their downstream brands. By comparison, independents thrive on small-scale production and entrepreneurial risk taking. Hence, while maturity and political hazards tend to drive global majors away, the same conditions in fact attract some independent players.

From a host country's perspective, therefore, increasing exploration and production levels involves a variety of options ranging from either shaping fiscal and political conditions to attract the *appropriate* types of outside players, or substituting outsiders through national operators and the occasional use of international service and technology providers. But why would any country set up its own import-competing NOC rather than create a regulatory and fiscal regime able to attract global specialists? The answer may lie in fundamental differences in how host countries and international operators view the attractiveness of a play or the necessary modes of operations and resource management. It may also reflect the perceived inability of any acceptable fiscal regime to effectively regulate international contractors.

Recently, host countries have moved in either direction. Whereas resource owners in the Gulf of Mexico, with the exception of Mexico itself, began to target global majors for large-scale deepwater investments, fiscal authorities in the mature North Sea focused on smaller, independent players. Similarly, while Russia's changes in licensing conditions and the transfer of Yukos' assets to Rosneft de facto renationalized resource access, Colombia separated its national oil company from that of the office of licensing to create more transparency and international investor interest. The results of all of this remain to be seen. Given that presently known OPEC and non-OPEC capacity expansions appear unlikely to close the impending production gap, which supply-side adjustments will be required and which are under way?

Part Two of this volume addresses issues of supply-side adjustments – from host countries' and IOCs' perspectives. Entering the perspective of resource owners, the first four tackle key elements of fiscal regimes, their evolution and geopolitical underpinnings.

David Johnston of Daniel Johnston & Co., Inc. outlines the characteristics of royalty/tax (R/T) systems, production sharing contracts (PSCs) and the composition of government take, which, at a given level of prospectivity, determine the attractiveness of a particular fiscal regime. What are the key elements to consider in designing a fiscal regime? What are the fundamental differences between negotiated licenses, as in Colombia and Indonesia, and different forms of fixed-term plus work-program/bonus bidding applied around the world? How did contracts evolve over the last hundred years and what are the most important global developments today?

Carlos Garibaldi, Managing Director of Harrison Lovegrove & Co. and in charge of their Houston office, reviews Latin America's macroeconomic and political context before benchmarking its petroleum economies in terms of perceived prospectivity and contractual and fiscal terms. The continent is a modest contributor to global oil and gas supply, with two globally relevant resource countries (Mexico and Venezuela) and several countries that happen to have hydrocarbons. With the exception of very few major plays, the majority presents interstitial niches for junior majors, international and indigenous independents, and transnational state oil companies (NOCs). After the perceived failure of liberalization policies of the 1990s, Garibaldi argues, the region has politically regressed in large part to an eclectic populist trend. In response to some of the stresses affecting the oil and gas industry (higher rents created by the demand-driven price windfall, resource psychology of scarcity, reserves concentration, petro-nationalism, company consolidation, NOC globalization, etc.) several local governments adapted their exploration and production (E&P) contractual and fiscal regimes to their new perception of the market. They seem to have chosen their positioning between bipolar directions, and their new stances may or may not be aligned with the industry's perceptions of their prospectivity and materiality.

In the North Sea, such alignment is vital. Gunnar Gjerde, Director General of the oil and gas department at the Norwegian Ministry of Petroleum and Energy, presents an overview of the history, present status and development plans for the Norwegian Continental Shelf (NCS) as a petroleum province. From the outset, creating value to benefit the Norwegian society, national management and control, building a Norwegian oil community and creating an attractive investment climate have been key goals for Norwegian oil and gas policies. Fundamental to all of this, the fiscal framework needed to be basically stable and yet offer authorities and investors sufficient flexibility to respond to changes in society, markets, industry trends and the development of the activities. Going forward, the Petroleum Directorate aims to add 5 billion barrels of reserves, exceeding current plans by about 1.5 billion barrels, which is the equivalent of one year of Norwegian total production at current record levels. This ambition comprises the most important issues looking ahead: adding reserves through exploration, getting discoveries on stream and adding value from existing assets through efficient operations. International investments are more than welcome. Quite different, it appears, from what currently may be observed in Russia.

Jennifer Josefson, a lawyer in the Energy Practice Group of White & Case's Moscow office, discusses the new draft for Russia's Subsoil Law, which was submitted by the government to the State Duma in June 2005. Driven by strong political and policy interest in reestablishing the government's hold on the Russian energy sector, the Draft Law impedes the ability of foreign investors to participate in the Russian upstream projects in general and particularly in relation to "deposits of strategic importance." Josefson offers an interesting discussion of the Russian definitions of subsoil ownership, subsoil use rights and the transferability of rights. While existing subsoil licenses are likely to remain in effect, only those subsoil users, which are currently Russian legal entities, will be entitled to convert their subsoil licenses into contracts at their discretion.

So what are the implications for IOCs going forward? How can they shape outcomes? The remaining chapters of Part Two try to address this. Álvaro F. Racero, Director North America of Repsol YPF, sets the stage by stressing the importance of reserve replacement in general, and particularly the interrelationship between "old" and "new" oil that is between exploration-based and recovery-efficiency based replacement. Since exploration alone has not been able to replace produced oil volumes since the early 1990s, worldwide oil reserves, in a broad sense, are currently decreasing. Racero argues that next to significant resources and capital, the oil and gas industry will need to embrace change if it is to provide the needed energy at the required pace and at affordable prices. But does this require, as it is sometimes argued, a move towards completely new business models?

Observers of current upstream oil and gas developments often call for but hardly ever reason the need for new business models based on an explanation of existing ones, their contexts and required modification. My own paper attempts to do just that. Chapter 7 describes types of upstream players – global majors, independents, contractors and technical service suppliers, NOCs and host country governments – and the characteristic challenges facing each of them. It then proceeds to offer a cursory assessment of the strategies pursued by BP, Spinnaker, Anadarko, Baker Energy, Petrofac, Halliburton and Schlumberger, as well as the UK government in managing the UKCS fiscal regime. Rather than pursuing new business models, each player seems to develop its given model in line with fairly conventional and even predictable patterns of industry evolution. Are advances in technology, economic and regulatory conditions likely to change this? Who is driving these changes and how can players prepare for it?

Aruna Subramanian, Strategy and Investment Analyst at Shell Technology Ventures, asks whether the future of technology in the exploration and production industry is doomed. Not only has the sector been traditionally much slower than other industries to accept innovation, but, over the past decade, oil majors have consistently reduced their R&D expenditure and relied on the service providers to come up with innovative solutions. Yet, service providers are hampered by ineffective risk-reward sharing contractual arrangements and pressures to commoditize new technologies and therefore are unlikely to come up with truly disruptive inventions. In addition, the long deployment process and complicated equity partnerships make it difficult for the small entrepreneur to survive. The chapter explores how collaborative ventures may be used to address these challenges.

The need for collaboration is also a crucial finding in the two final papers of this section. Olav Ljosne, Regional External Affairs Manager for Shell Exploration and Production Africa, outlines Shell's most recent scenario thinking. He asks how, given complex and highly uncertain geopolitical supply constraints, corporate strategies can be based on robust reasoning. Next to some predetermined trends, such as the changing role of China and growing concerns with urbanization and greenhouse gases, Shell's global scenarios for 2025 see the global business environment as being shaped by three main factors: the impact or reliance on market incentives, central regulation, and social cohesion. Cast as a trilemma, Ljosne outlines the operational implications emerging from three scenarios: "low trust globalization," "open doors," and "flags." Going forward, collective actions involving a range of stakeholders will replace ineffective stand-alone strategies.

Complementing and contrasting Olav Ljosne's macro perspective, Christine Batruch, Vice President Corporate Responsibility, Lundin Petroleum AB, details a fascinating account of public affairs and stakeholder management. Lundin Petroleum obtained the rights to explore for and produce oil and gas in concession Block 5A, Unity State, Sudan, in February 1997; it sold these rights in June 2003. During the period in which the company was active in Sudan, it operated in the belief that oil could benefit the economic development of the area and the country as a whole, and that this would have a catalyzing effect on the peace process. The problems that it encountered in the area, however, led the company to constantly reassess its activities, role and responsibilities there. This chapter examines the reasons why Lundin decided to operate in Sudan, the challenges it faced in the course of its activities, the steps it adopted to satisfy both its commercial objectives and ethical concerns, and its efforts to promote a peaceful resolution of the conflict.

Clearly, supply-side changes to increase production and refining capacities and to enhance exploration and recovery efficiency tie resource owners, and national and international oil companies into an intricate web of market and non-market coordination. The process meshes national resource management objectives and fiscal policy targets with diverse corporate and bureaucratic agendas and geopolitical interests. Its complexity defies simple predictions about potential energy futures, the value of which would be anyway questionable if not controlled for potential demand-side adjustments. Focusing on these, the next section first sketches the demand-side impact of oil price increases as well as opportunities for fuel conservation and substitution, before outlining related contributions of conference participants.

Demand-side adjustments – from macro to micro

Oil price increases affect countries differently depending on the energy intensity of their output and the availability of domestic buffers to cushion them. Whereas in 2005, increased energy costs added a mere 1.3% to US core inflation and 2.3% to German producer prices, emerging economies suffered more from oil-price hikes. India's economic output, for instance, is nearly three times as energy intensive as the average OECD production; China's ratio of oil use to GDP is almost twice that of the USA, which in turn, even after halving its energy intensity over the last three decades, remains at double the Japanese level. Facing an apparent trade-off between rapid growth and lengthy adjustment, China, at the time of writing, continues to opt for fuel price controls rather than curbing demand.

By contrast, more energy efficient countries often delay adjustments in view of stakeholder interests rather than the need to grow. As a result, economies with competitive and deregulated energy consuming sectors better contain cost-push inflation than those, in which, for instance,

Table 1.2 Hydrocarbon Net-Importers and Exposure

Information as of 2003	Oil production Mtoe	Oil consumption Mtoe	Imbalance Mtoe	Actual net oil imports Mtoe	Energy intensity[a] toe/'000$	Oil as % of TPES %	Government tax as % of fuel price[b]
USA	351	848	(497.34)	595	0.2208	40.4	15.2
EU	305	626	(321.08)	437	0.1620	37.6	55.5
Japan	1	219	(218.29)	261	0.1521	49.7	48.5
China	170	218	(47.93)	106	0.2300	19.2	33.8
India	39	103	(64.73)	85	0.1900	22.4	38.4

Mtoe = million tonnes of oil equivalent TPES = Total Primary Energy Supply

[a] Energy intensity (defined as primary energy to generate a unit of GDP) = TPES/GDP, (toe per thousand 2000 US$ PPP)

[b] Unleaded Regular gasoline except India which is for Regular Leaded gasoline. For EU, the median tax is given.

Source: Adapted from IEA/OECD most recent data

inadequate regulation permits utilities to pass on oil price increases to electricity consumers – even if their fuel mix is not overly hydrocarbon-based. The same is true for cases where governments' windfall tax gains, particularly at the oil pump, are transparent rather than hidden from the voting public. Clearly, the prospects for fuel conservation and substitution are related to more than the costs of oil consumption and the benefit of switching. Effective demand-side management means sorting out competing stakeholder interests. The question is how?

Consider the fate of public and private initiatives aimed at slowing global warming. International policy ventures range from the Kyoto CO_2 and greenhouse gases (GHG) reduction targets to the EU's Externe Projects to quantify the total costs of energy use; technology projects seek to improve combustion engines, drive up the level of co-generation or push for hydrogen and fuel-cell applications; corporate endeavors set up hydrogen infrastructures, offer software for intermittent power supply or improve facilities for energy storage. In each case, success requires entire supply chains to switch behavior and calls for a mix of market and regulatory incentives. But markets often fail, policies may be ill-conceived and coordination blocked by strategic games and political strife.

For instance, EU member states translated Kyoto reduction targets into renewable obligations for electricity suppliers. Yet, in countries like the UK or Germany, consumer enthusiasm following the announcement of green electricity hardly ever translated into sales. To quote the CEO of a leading European utility, "[c]onsumers applaud green initiatives but are unwilling to pay for them." At the business-to-business level, market-based adjustments are not necessarily easier. As of 2005, emissions trading in the EU should create financial incentives to shift towards renewable resources. But although the trading prices for CO_2 allowances stimulate the use of existing fuel mixes in cleaner and more efficient ways, they are still too low to make unsubsidized onshore or offshore wind or photovoltaic production competitive. Do policies need to set reference prices to more adequately reflect the value of avoided CO_2 pollution?

But policymakers face their own set of challenges. Ponder two cases of many. First, various European utilities are able to produce nuclear electricity at around 40% of the cost of producing a kilowatt-hour with natural gas and coal respectively; doing so could also triple their effective earnings if carbon taxes were included. In addition, a completely non-nuclear EU that otherwise maintains its current fuel mix would see its GHG emission go up by 55%. And yet, the EU's non-nuclear lobby, particularly Italy, Germany, Austria and Belgium, is doing everything to block attempts by countries like Finland to increase their nuclear capacity and diversify their fuel mix. But is it at all reasonable to expect effective policy coordination in the absence of a common European energy policy?

Second, although road, domestic and international air transports are by far the biggest and fastest growing contributors to CO_2 emissions worldwide, the Kyoto Protocol conveniently stayed away from addressing the most sensitive parts of the sector. In the absence of truly binding, global commitments, national and regional authorities are acutely aware of the political risk they take when filling Kyoto-gaps with measures to curb emissions, set consumption targets or increase transport efficiency. But political concerns need not limit adjustment for everyone. In China, during the Olympic Games in 2008, all cars directly or indirectly linked to the event will drive with synthetics or biofuels. In Europe, Airbus, aware that by 2050 emission from growing regional air transport is projected to eliminate all of the EU's CO_2 reductions achieved by then, is heading the CRYOPLANE project to develop hydrogen for use in aviation.

Clearly, even if short-termism were to limit the chance for decentralized fuel substitution – either through markets or democratic processes – more sheltered political or corporate planning may still be able to set the long-term agenda. The question is what would motivate such visionary, albeit not immediately lucrative, behavior? Would redistribution be required to initiate or legitimize outcomes? Which institutional context would help to build consensus at national and international levels? Part Three of this volume endeavors to answer some of these questions.

David R. Demers, CEO of Westport Innovations Inc., focuses on developing fuel efficient transport technologies. His chapter looks at the interplay between oil price, environmental damage and the growth of the transportation sector in Asia with a focus on China. In the face of apparent supply constraints and severe economic disruptions, he argues, there is a need to immediately bleed off some of the pressure mainly stemming from the transportation sector. This could be done by either rationing, price-induced conservation, or, as in China and India, also by accelerating the use of alternative fuels. Based on a review of alternative fuels ranging from biofuels, and hydrogen to methane, Demers argues that natural gas will pave the way for more advanced gaseous fuels such as hydrogen. It not only proves to be an excellent vehicle fuel in many parts of the world, but large investment in liquefied natural gas (LNG) facilities may offer China a unique window to change its transportation sector and thereby the global medium-term oil demand outlook.

While Demers proposes that hydrogen technologies will be taking over in the long run, current progress towards the hydrogen economy evokes

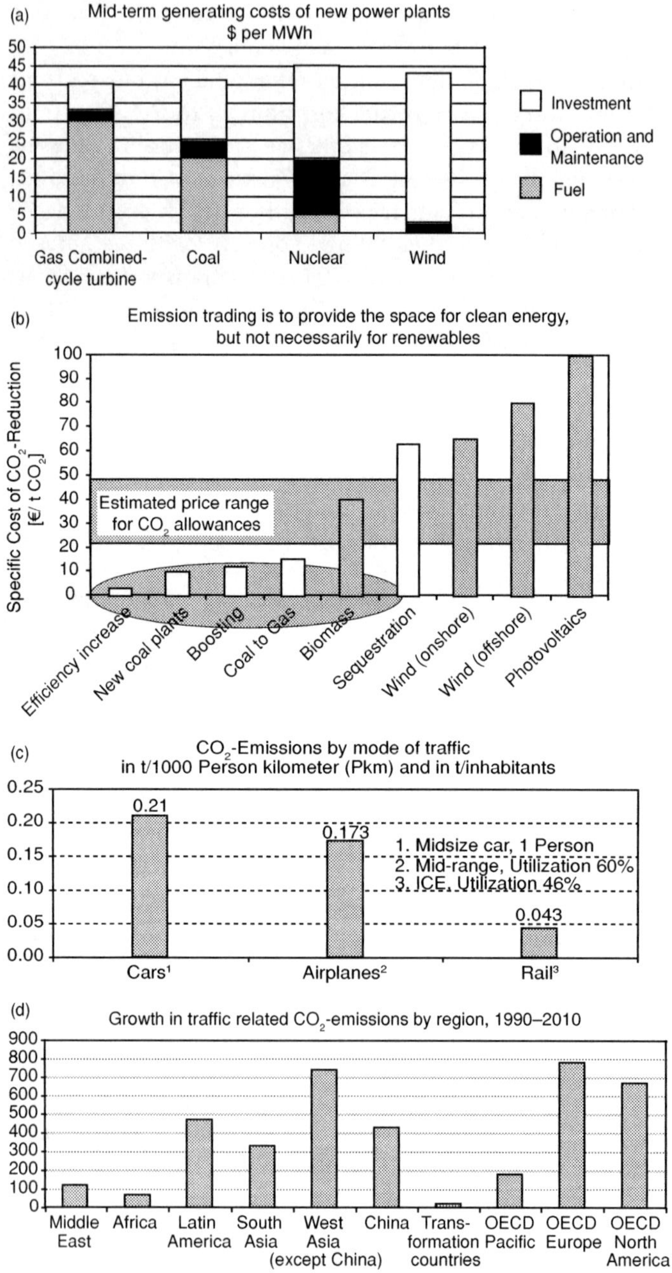

Figure 1.3 Fuel Substitution – Costs and Impact[6]

the impression of a visionary mirage. What institutional conditions are required to move ahead and are there any signs of a significant advance? Addressing these concerns, Karen Hall, Vice President of Technical Operations for the National Hydrogen Association, USA, outlines the position and activities of this US industry body linking more than 70 companies and organizations to promote the use of hydrogen energy and technology. For hydrogen to deliver towards objectives such as national energy security, environmental stewardship and economic prosperity, standards, institutions and technological and commercial breakthroughs are required – each case calls for coordination and consensus building. Public–private interaction is needed to effect fuel substitution. The next three chapters deal with this and the interaction of economic regulation, environmental standards and corporate strategies in the area of electricity supply.

Caryl Pfeiffer, Director, Corporate Fuels and By-Products, LG&E Energy Corporation, Kentucky, offers a perspective on how the Clean Air Interstate Rule will affect investment and management decisions in the US electricity sector. Following an introduction to the US Acid Rain Cap and Trade Program, its regulatory drivers and political subtleties, Pfeiffer discusses compliance options including investing in air pollution control technologies, lowering emitting generation sources, switching fuel or relying on the emissions trading market. The chapter finishes with a discussion of the business case of LG&E's own compliance program and its fit with the company's overall strategy.

Bjørn N. Holsen, Sergio Castedo and Geir Holler address the strategic intention of Norway's Statkraft to invest in environmentally-friendly generation capacity under uncertain regulatory conditions. Low rates of capacity investments in Scandinavia since deregulation brought a then existing surplus capacity down; today there is a growing concern over the ability of market-based power systems to generate sufficient and timely investments. Companies must consider the effects of high fuel prices, Kyoto objectives, emission trading schemes and renewable targets. As gas prices have increased more than coal prices, it has become more costly to alter the merit order between coal- and gas-fired power plants. Hence, while the role of natural gas as the fuel of choice for base-load power generation is increasingly called into question, the cost of CO_2 emissions reductions is now higher for the same reduction target. Although Statkraft is focusing on the development of new capacity in natural gas combined cycle generation, wind and hydropower, the lack of green incentives in Norway has made renewables, by and large, an underutilized resource. At the European level there seem to be no clear long-term risk-reward mechanisms required for new investments. As a result, governments' guidelines will be critical for fuel choice.

Stefan Bockamp and Jörg Kruhl, respectively Coordinator Emissions Trading and Head of Technology Policy, E.ON Energie AG, echo this view. Their chapter reflects upon E.ON's experience with the European Emissions Trading Scheme, launched on January 1, 2005. Emissions-trading affects the entire business system of electricity supply as well as investments going forward. Bockamp and Kruhl point to conflicting interpretations of standards, procedures and responsibilities, their impact on markets and operations and suggest modifications to enhance the transparency, stability and neutrality of outcomes for the upcoming, second trading period. In their view, there is not only a need to strike a balance between the three equally important objectives of economic viability, energy security and environmental protection, but, particularly with regard to the latter, a requirement to coordinate objectives and policies at the global level.

Clearly, Bockamp and Kruhl's plea may be tied to the spirit of the Clean Development Initiatives (CDIs) and the broader Kyoto endeavors, but policy concerns are not surprisingly very different if seen from an explicitly global and developmental perspective. Assuming this vantage point, an ever-growing body of literature – from early World Bank proposals to the recently published *Limits to Growth: The Thirty Year Update* – makes a case for drastically reorienting energy policies and global technology sharing.[7] The final chapter of this collection goes beyond that. Subhes C. Bhattacharyya, Senior Lecturer at the Centre for Energy, Petroleum, and Mineral Law and Policy (CEPMLP), University of Dundee, argues for a fundamental policy shift. With symptoms of unsustainable energy practices abounding and a history of sustained market failures influencing the sector's policies, it is obvious that the energy sector requires a return to a sustainable path. Policies involving the three Rs of the 1990s (right prices, reform and renewable energies) fail to yield desired results. The chapter calls for strategies to provide energy access to the energy have-nots of the world.

Energy futures – the meeting and outlook

Bringing together energy experts of rather different analytical views, doctrinal convictions and stakeholder affiliation, IMD's research conference stimulated an open discourse on the fundamental demand-side and supply-side challenges that shape our energy future. Papers had been commissioned, arranged and edited to link a broad range of topics. No attempt was made to present one unifying, comprehensive outlook. Nevertheless, discussions in plenum and subgroups often gravitated towards a common theme – the flaw, shared by many current energy scenarios, to mix normative and positive perspectives so as to conveniently depict energy futures as either primarily market or policy driven. Consider the case of the CIEP study on energy supply securities and geopolitics, prepared for the EU Commission's DG Energy and Transport in 2004.[8]

CIEP offers two scenarios: in *Markets and Institutions*, energy markets continue to globalize, and political and economic institutions as well as policies and strategies contribute to a rather smooth, market-driven expansion of production and distribution capacities, viz. a reduction of GHG emissions. As markets deepen, national and regional demand and crisis management strategies lose in importance. Conversely, in *Regions and Empires*, failures of political and economic institutions to deal with trade, environmental and Middle-Eastern security concerns exacerbate competition for energy access and severely weaken transatlantic relations. Energy security overrides arguments for market efficiency, promotes bilateral accords at the expense of multilateral trade and ultimately dampens production and capacity expansion. In view of developments in the UN Security Council, WTO and the Kyoto process, CIEP concludes that the world's current state is more in line with the second scenario and predicts an increase in protectionist energy policies and geopolitical adjustments, particularly in the Caspian region and the Far East.

Of course, it is not the purpose of scenarios to assess whether something will happen but rather to ask what one would do if it were to happen. Still, in distinguishing conceivable, feasible and likely states of the world, fundamental references need to be clear. The problem is that with regard to energy, "markets" and "geopolitics" do not offer such clear-cut, mutually exclusive orientations.

For one, energy markets are hardly ever free; they are at times not even set up to value resources properly and are often framed to pursue short-term objectives irrespective of long-term consequences. Different from renewables, hydrocarbons, for instance, are part of a country's finite capital endowment – a fact that is often neglected in their valuation. While the Middle East is depleting its rich but finite oil reserves, Italy and Spain depend on more than 85% and 78% of imported hydrocarbons respectively; neither one makes sufficient use of its renewable solar potential. Early US policies to promote its domestic industry ultimately meant that its own oil resources were exploited earliest and at relatively low prices; current EU energy liberalization is to lower consumer prices and promote industrial competitiveness but thereby is liable to impede energy efficiency and conservation. While linking greenhouse gas emissions and global warming[9] is increasingly accepted, reminders of economic consequences are rather rare: According to UNEP, the UN Environment Programme, Europe suffered commercial losses of US$33 billion in 2002 as a result of adverse weather. Clearly, energy markets that treat renewables and non-renewables alike and fail to internalize production externalities cannot but prompt inadequate production and consumption decisions. How then can they provide the bedrock for any "optimistic scenario"?

Similarly, it is often held that geopolitics not markets are driving energy outlooks. Beginning with the early 1970s, two oil shocks fundamentally transformed the modern history of energy. Pricing power shifted from the Texas Railroad Commission to OPEC whose price hikes set the stage for a subsequent dramatic fall in price and in the cartel's market share. Going forward, OPEC decisions will regain prominence, large consuming countries, such as China and India, will seek to lock up new energy resources worldwide, and others may attempt to politically coordinate a shift towards non-hydrocarbon substitutes. Either way, however, market alternatives and economic needs will continue to limit any political discretion. With Saudi Arabia's income per capita falling continuously since 1980, the Kingdom must balance OPEC objectives with the growing need to provide for basic social services. Both Venezuela's political rhetoric and its welfare programs in the end depend on international infrastructure investments. Support for Kyoto-type processes will be assessed against costs and notions of competitiveness. Geopolitics meets the market-test.

Clearly, assessing and managing our energy future calls for substituting stylized "market" vs "statist" macro-perspectives with micro-analyses of resource economics, environmental impact and the reality of market and non-market decision-making. Separating "is" from "ought" and focusing on the former avoids the trap of naïve optimism or stifling pessimism and allows for a constructive search for realistic solutions to current energy challenges. It is the common denominator shared by all the contributions in this volume.

Notes

1. IEA, *World Energy Outlook*, Paris, 2005.
2. According to the *Oil and Gas Journal*, of the world's 1.28 trillion barrels of proven reserves, 885 billion barrels (69%) are held by OPEC, as of January 2005. The non-OPEC reserves include Canadian non-conventional reserves. Not including

Canada, the world's proven oil reserves are about 1.1 trillion barrels, of which OPEC holds 84%.

3. In February 2006, OPEC cited "concern over rising tensions in the Middle East, a 120,000 barrels per day disruption in Nigerian supply and a cold snap in Europe, which iced in 250,000 b/d in Russian output" to reason continued upward movements in OPEC basket prices. OPEC, *Monthly Oil Market Report*, Vienna, 2006.
4. IMF, *Oil Market Developments and Issues*, March 1, 2005.
5. According to the US Department of Energy and IMF the total is 1212.9 billion barrels (excluding Canadian oil sands).
6. VCD (Verkehrsklub Deutschland), *Zügig durch Europa*, Bonn, 2003; WBGU (Wissenschaftlicher Beirat der Bundesregierung Globale Umweltveränderungen), *Welt im Wandel: Energiewende zur Nachhaltigkeit*, Berlin, 2003.
7. See, for example, World Bank, *Fuel for Thought: An Environmental Strategy for the Energy Sector*, Washington, D.C., 1999; D. Meadows et al., *Limits to Growth: The 30 Year Update*, London: Earthscan, 2005.
8. Clingendael International Energy Program (CIEP), *Study on Energy Supply Security and Geopolitics*, report prepared for the EU Commission DGTREN, 2004.
9. The global average land and sea surface temperatures in May 2003 were the second highest since records began in 1880, and the highest if land temperatures only are taken into account. See A. Webb, "The Relationship between Energy and Environment," *European Policy Analysts*, Economist Intelligence Unit, 2003.

Part Two

Upstream

2
Petroleum Fiscal Systems: Royalty/Tax Systems, Production Sharing Contracts and Service Agreements Compared

David Johnston

The design of a petroleum fiscal system is much more important than the type of system used. Governments can achieve their fiscal objectives with whichever fiscal system they choose as long as the system is designed properly. Improper design can translate into money-left-on-the-table. Philosophical and political objectives of course are another matter. The *type* of system used is much more important to the international oil companies (IOCs). On the one hand, the IOCs would prefer to work with a fiscal system that allows them to "book more barrels" but they also want a stable environment. Usually a royalty/tax (R/T) system will allow them to book more barrels but a production sharing contract (PSC) can provide more stability in many cases. The IOCs are concerned with fiscal stability no doubt, and who can blame them? But, more and more these days, stability is less a function of fiscal system type, than design.

Different governments have different needs. All governments have different boundary conditions, concerns and objectives. However they are almost all interested in enhancing exploration activity.

There are also fundamental differences when it comes to exploration acreage or projects on the one hand and non-exploration projects (development or enhanced oil recovery projects (EOR)) on the other. When it comes to exploration, the division of profits (take) is a central focus. With development or EOR projects though, the focus is more on internal rate of return (IRR). The former, much riskier for the IOC, must provide a chance for the IOC to benefit from the up-side in the event of a discovery. The latter, with less risk, should hopefully provide a fair return on investment.

Beyond the division of profits is another issue. When governments do not receive a reasonable percentage of revenue (or production) each and every accounting period, referred to as the effective royalty rate (ERR), an unhealthy business environment can be expected.

The selection of system type is important for countries trying to attract IOCs for exploration activity, particularly if prospectivity is not good. The means by which governments allocate acreage or projects is on par with the importance of the proper fiscal design.

Introduction

Oil is the world's number one strategic commodity. It is of vital interest to developed and developing nations that rely more and more on imported oil and gas. It is also vitally important to exporting nations, many of them some of the poorest countries in the world – the Middle East aside. For countries with petroleum resources the contribution from the petroleum sector to the nation's budget is often dramatically greater than the contribution to the country's gross national product (GNP). For example if the petroleum sector were to represent say 10% of GNP it would likely represent from 30% to 40% of the nation's budget. There are two reasons for this: first, relative to most other industries petroleum is very profitable; second, the effective tax rate (government take) for the petroleum industry relative to other industries is usually about double or more.

Numerous dynamics influence today's industry. Oil demand continues to grow, and at a faster rate than was expected. Consumption went from 79 million barrels a day in 2002 to 84.5 million in 2004, leaping by 2 to 3 million barrels of oil per day (BOPD) each year when expectations had been on the order of 1 to 1.5 million BOPD growth per year. Much of the new demand comes from the Asian giants India and China. But supply of oil and gas is a function of exploration and production. There is now every indication that exploration and the resultant discoveries, have "peaked." We do not know for sure when production will peak, as production peaks lag behind exploration peaks, sometimes by as much as 30 years. Gas is becoming more and more important but gas discoveries are still characterized (somewhat facetiously) as "worse than a dry hole" in many regions of the world. That day and age is fast disappearing, although the fact remains that gas is simply much more difficult to transport than oil and is still flared in many parts of the world – nearly 10 billion cubic feet of gas per day. Nigeria flares almost 2 billion cubic feet per day in their Niger Delta oil fields – not far from some of the poorest people in the world. And in many other parts of the world gas discoveries are simply "shut in."

As these features change there are also changes in relations between the main players in the industry: the countries and national oil companies (NOCs) that control the bulk of the available oil and gas reserves, and the international oil companies (IOCs) that provide much of the financial, technical, organizational and marketing needs of exporting and importing countries. On the side of the producing countries there are economic and political complexities associated with managing oil and gas. These issues are important for countries, but also affect the way they relate to private oil companies. Many of the problems associated with oil and gas exploration and production, particularly in low-income countries, can be associated with corruption. But in some cases the problems stem from misunderstandings and poor communication. In these cases, the government, NOC and the IOCs are suspected of stealing a nation's wealth. Indigenous peoples of the world no longer sit idly by. The results are usually not healthy.

While relations may be fraught with political difficulties, they are also important from a practical point of view. There is increasing competition among countries for the limited resources of the IOCs. The ability of countries to attract IOC investment depends on prospectivity, stability and their marketing skills. IOCs want to reduce risk in an inherently risky business.

This is the subject of this chapter: how do governments, NOCs and IOCs work together and what types of contractual relations are likely to lead to better outcomes than we have seen in the past? The question is examined by focusing more on the elements of an agreement than on the broad differences between the "families" or types of systems that exist. The reason is, although there are myriad ways to structure business relationships in the petroleum sector, the first observation is that in fact, for all practical purposes, *there are only two main families of Petroleum Fiscal Regime.*

But, strikingly, from a financial point of view, the similarities within and between these two families dramatically outweigh the differences. For the most part, R/T systems and PSCs can be designed to provide identical financial results for all parties.

Comparative analysis: royalty/tax systems – production sharing contracts – service agreements

A thorough understanding of any type of petroleum fiscal system requires a clear vision of its entire context – i.e. an understanding of all systems.

From a mechanical point of view there are practically no differences between the various systems. The hierarchy of arithmetic such as (1) generation of production and revenue followed by (2) royalty or royalty equivalent elements, followed by (3) cost recovery, tax deductions or reimbursement etc. and (4) profits-based mechanisms such as profit-oil sharing and/or taxes is found in almost all systems. There are some interesting exceptions to this general rule and they are most likely to be found among the service agreements of this world.

There are two main families of fiscal system: "concessionary" systems, more commonly known these days as royalty/tax (R/T) systems, and the "contractual based" systems, which include both production sharing contracts (PSCs) and service agreements (SAs). There are some risk service agreements that appear to have more of the characteristics of a royalty/tax system – Venezuela (with royalties and taxes), and some look more like a PSC – Philippines (with a cost recovery limit and profit oil split). *The distinguishing characteristic of each is where, when, and if ownership of the hydrocarbons transfers to the international oil company.* Numerous variations and twists are found under both the royalty/tax (concessionary) systems and the contractual-based themes. The taxonomy of petroleum fiscal systems is outlined in Figure 2.1.

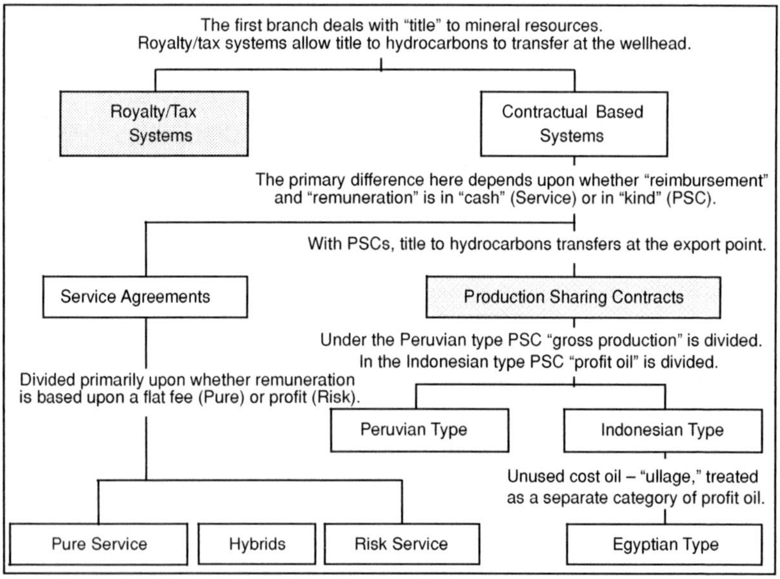

Source: Daniel Johnston & Co., Inc.

Figure 2.1 Classification of Petroleum Fiscal Regimes

In fact, preferences for one system over another and certain elements or conventions generally tend to be regional. Hence R/T systems are often simply referred to as concessions in some areas. But, the term "concession" has negative connotations in other parts of the world. It is simply not so politically correct. Political correctness also influences the difference between "PSC" and "PSA" terminology. In Russia, the word "agreement" is favored over the word "contract" because of the negative connotation of the word "contract" when translated into Russian. Yet, a PSA is the same thing as a PSC. Some of the geographic influences can be seen in Table 2.1.

Table 2.1 Types of Agreements and Where They are Used

Type of Agreement	Region
Service agreements	Latin America and M. East
Rate-of-return (ROR) features	Africa and Former Soviet Union
C/R limits based on net production	Africa
Work program bidding	Former British Colonies
PSA terminology (vs PSC)	Former Soviet Union
"Cost stop" terminology (vs C/R limit)	W. Africa
Taxes "in lieu"	Middle East

The belief that systems are somehow fundamentally different from a financial point of view has led to a number of common misconceptions. Despite common claims to the contrary, neither R/T systems nor PSCs are inherently more likely to allocate more risk to either the NOC or the IOC. Similarly, it is not the case that PSCs "allow the IOCs to get their costs back faster" or even that they get them back at all. Nor is it necessarily true that PSCs are more stable than R/T systems or that PSAs (PSCs) are unsustainable.

In order to work out the merits of a particular agreement, most easy rules of thumb are unreliable; rather, a deeper understanding of how the different systems operate and particularly the core fiscal elements is more meaningful.

The first step then is to generate a common understanding of the different systems. This unfortunately is rendered difficult by the *lack of standardized terminology*. On the *technical* side of oil exploration and production, terms like porosity, permeability, API gravity, pressure and saturation are universally used and understood. *This is not the case on the financial/fiscal side of the business. Comparing fiscal systems is not facilitated by standard well-established terminology.*

In response, the following section of this chapter begins by providing a working vocabulary and a working understanding of the two primary fiscal systems used in the world today. It starts with a discussion of the concerns facing governments and examines the notion of "government take." Government take is the most predominantly cited statistic, and certainly an important one; but while commonly used, the government take statistic is flawed. For this reason companion statistics and other features of agreements are examined. In particular we address the questions: (1) what is the "effective royalty rate"? (2) what is the bonus structure (if there is one)? (3) how does the agreement reflect changing circumstances, and in particular, if oil prices rise or fall how does the percentage going to the government change? (4) are there (many) "crypto" taxes? and (5) does the agreement involve government participation?

Also it is important to governments just *who* the IOC is. Such features matter for ensuring that the right kind of IOC is attracted to the prospect and for getting the incentives right once they begin operations. The third section of this chapter examines: (1) the incentives the contractor has to keep costs down, (2) relinquishment, (3) the domestic market obligations, (4) what percentage of the production is the contractor "entitled" to lift and what percentage of the reserves can the contractor "book"? (5) the stability of the regime, and (6) whether, and how, the prospect is ringfenced.

The fourth section turns to empirical aspects, examining how these different features have played out in different parts of the world: what kind of structure has been successful, where, and why. The final section highlights the core messages from the analysis and suggests fruitful avenues for future research.

Government concerns

With the exception of the United States, Canada, and a very few old Spanish land grants in Colombia, mineral rights belong to the state. And, in most countries the nation's mineral wealth is considered a "Gift from God." The result is that managing a country's mineral wealth is considered a "sacred trust" even though, in many situations, the nation's mineral wealth benefits only a few people.

In short, countries with limited proven mineral wealth are seeking exploration activity and have limited leeway attracting it. Still, they want the best "terms" they can get. All countries have their own unique boundary conditions, concerns and objectives. And, needs, traditions, perspectives, perceptions and politics differ as well. But in general the major concerns facing a country are:

Table 2.2 Typical Contract Conditions

Area	Block sizes range from extremely small for development/EOR projects to very large blocks for exploration. Typical exploration block sizes are in the order of 250,000 acres (1,000 km^2) to over a million acres (>4,000 km^2).
Duration	Exploration – Typically 3 Phases totaling 6 to 8 years. Production – 20 to 30 years (typically at least 25 years).
Relinquishment	Exploration 25% after 1st phase, 25% of "original" area after 2nd phase. This is most common but there is wide variation.
Exploration obligations	Includes seismic data acquisition and drilling. Sometimes contract requirements can be very aggressive in terms of money and timing – depends on the situation. All blocks are different.
Royalty	World average is around 7%. Most systems either have a royalty or an effective royalty (ERR) due to the effect of a cost recovery limit.
Profit oil split	Unique to PSCs and some service agreements. Most profit oil splits (approximately 55–60%) are based upon a production-based sliding scale. Others (around 20–25%) are based upon an "R" factor or ROR system.
Cost recovery limit	Unique to PSCs and some service agreements. Average 65%. Typically PSCs have a limit and most are based on gross revenues. Some (perhaps around 20%) are based on net production or net revenues (net of royalty). Over 20% have no limit (i.e. 100%). Approximately half of the PSCs in the world have no depreciation for cost recovery purposes – but almost all do for tax calculation purposes.
Taxation	World average Corporate Income Tax (CIT) is probably between 30% and 35%. However, many PSCs have taxes paid "in lieu" – "for and on behalf of the contractor" out of NOC share of profit oil.
Depreciation	World average is five-year straight line decline (SLD) for capital costs. Usually depreciation begins "when placed in service" or "when production begins" whichever occurs later.
Ringfencing	Most countries erect a "ringfence" (55%) or a modified ringfence (13%) around the contract area and do not allow costs from one block to be recovered from another, nor do they allow costs to "cross the fence" for tax calculation purposes.
Government participation	Typically the NOC (or equivalent) is "carried" through exploration. Approximately half of the countries with the option to participate do not reimburse "past costs."
Effective royalty rate	The minimum government take for a given accounting period.
Crypto taxes	Crypto taxes are those costs and obligations the contractor must take on that are not readily captured in the take calculations.

Source: D. Johnston, *International Petroleum Fiscal Systems*, PennWell Books, 2001

1. Control over the country's resources
2. The need to *attract* investment and to attract the right *kind* of company, even if the financial conditions are not as good. (Trinidad awarded a block to BHP even though Talisman submitted a higher bid. The government was familiar and comfortable with BHP.)
3. Getting a large (and fair) share of the profits (take) while keeping costs down
4. Guaranteeing a certain share *each accounting period* (effective royalty rate and/or minimum government take)
5. Maximum efficient production rate (MEPR)

Control

Contract evolution is driven by governments demanding greater control of resources, countered by oil companies pushing for stability.

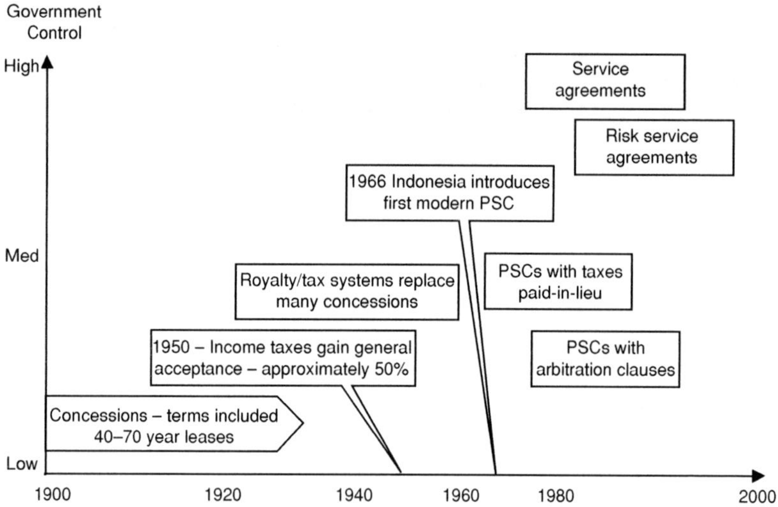

Source: Daniel Johnston & Co., Inc.

Figure 2.2 Contract Evolution and Government Control

The "government take" statistic

Government take represents the government share of economic profits from almost all means by which the state extracts rent: bonuses, royalties, profit oil, taxes, government working interest, and so on. By definition crypto taxes are rarely adequately captured in the take statistics. Because of the dynamics of bonuses, royalties, cost recovery limits, sliding scales and various other mechanisms, there can be various levels of government take for any given system depending upon a variety of circumstances.

Table 2.3 Government Take – The Common Denominator

		Take definitions
Economic profit ($)	=	Cumulative gross revenues less cumulative gross costs over life of the project (full cycle). Also referred to as cash flow.
Government take (%)	=	Government receipts from royalties, taxes, bonuses, production or profit sharing, and government participation, divided by total economic profit
Contractor take (%)	=	1 – Government take
	=	Contractor net cash flow divided by economic profit
Company take (%)	=	1 – Government take (excluding government participation)
	=	Company net cash flow divided by economic profit

Notes: (1) Here a distinction is made between contractor (or more precisely contractor group) and an individual company within the contractor group.
(2) In the past, most "take" statistics were based upon undiscounted cash flow. More recently take statistics are being quoted from a present value point of view (i.e. the division of discounted cash flow).

Government take is one of the most important statistics used in the oil industry when it comes to fiscal system analysis or design. It is important to remember that the government take statistic represents the division of profits "full cycle" over the full life of a field or fields. Government take statistics that represent something less than "full cycle" analysis are rare and for good reason – they usually provide little or no analytical insight.

After Bolivia's First Gas War in 2003, a new fiscal system was proposed. It would increase the share of revenue to the Bolivian government in the early years of production from their newly discovered gas fields. The proposed system left government take virtually unchanged but meant that the government would get its share of profits earlier. It also attempted to keep the revolutionaries happy without completely alienating the oil companies that risked capital exploring for and finding Bolivia's vast gas reservoirs.

Bolivia needed money sooner rather than later. But a comparison of the proposed system with the previously designed systems using undiscounted government take would not have shown a difference. Bolivia ultimately has proposed to increase government take radically with virtually a 50% royalty with much rhetoric about greedy oil companies.

Few politicians or citizens around the world appreciate or give sufficient credit to oil companies for the risks they undertake in petroleum exploration.

Few developing countries are able or willing to wait for profits to be generated from a developing field before they get a share. That is why we see signature bonuses and other front-end-loaded elements like royalties and cost recovery limits.

A front-end-loaded system provides more money to a government in early years. However, government take statistics alone rarely indicate if the system is front-end-loaded. That is why companion statistics help so much. Effective royalty rate is a companion statistic that helps show "how" a government takes.

Below is a list of weaknesses in the government take statistic:

- While the government take statistic is important it is only one of about 12 metrics and elements that help characterize and quantify the essence of a system.
- Does not adequately capture signature bonuses.
- Often does not address present value and almost never factors-in risk.
- Does not address "how" government takes (such as front-end-loading).
- The companion statistic "ERR" helps here.
- Says nothing of timing and time value of money (unless "discounted").
- Its macroeconomic scope is too narrow – it does not measure *all* of the means by which a government benefits, i.e. gross benefits.
- Says nothing of ringfencing.
- Does not measure contract or system stability.
- Reserve/lifting entitlements and "ownership" is not accounted for.
- Does not differentiate between diverse work program provisions.
- By definition "crypto taxes" don't get captured.
- It is not relevant in some important situations. Government take for exploration may not be the same statistic for development (i.e. the government participation controversy).

Country fiscal systems are often compared to those of neighboring countries. Chad, for example, is projected to retain an average of 28% of the oil value for the period 2002–10. Meanwhile, Nigeria, Gabon,

Cameroon, Congo-Brazzaville, Angola and Equatorial Guinea retain a mean 80%, 75%, 75%, 66%, 60% and 60% respectively.[1] Unfortunately this particular comparison is misleading. The low rate for Chad is ascribed to a high degree of political risk in Chad and Cameroon, the low quality of crude oil, and the fact that an expensive pipeline had to be constructed in order to export Chad's crude oil. The comparison was also used to support the thesis that "Chad's inexperience in negotiating with one of the world's largest companies should not be overlooked as an important factor [in Chad's low take]."[2]

One curious thing to note with the above comparison is the time frame used, 2002–10. Since Chad oil did not start shipping until 2003, this time frame represents only the early years of production when taxes would be minimal. It suggests that the take calculation has not been "full cycle." If this is the case, the comparison above is probably more of a representation of Chad's ERR than government take.

To consider another example, plotting government take against project profitability (Figure 2.3) points out a key difference between the R/T system, and PSC compared earlier.

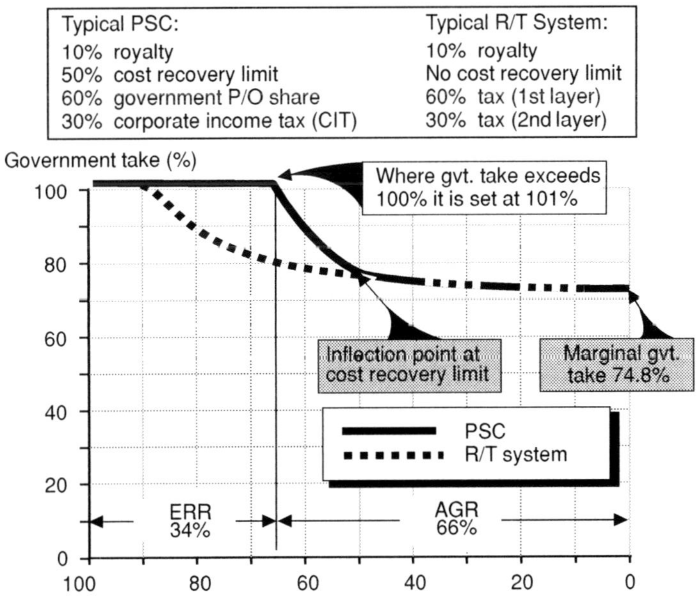

Source: Daniel Johnston & Co., Inc.

Figure 2.3 Government Take vs Project Profitability

Government take is much higher for a PSC with low profitability projects. This graph shows how the PSC in this particular case is more front-end-loaded than the example R/T system. It is the cost recovery limit that makes the PSC more front-end-loaded (or regressive) than the R/T system.

In early years, government revenue is guaranteed for both systems because of the royalty. The PSC however also has the cost recovery limit that guarantees the government additional revenue. In fact the government take for sub-marginal fields can be extremely high, so graphs like this are usually capped at 101% – showing takes in the 100%+ range is relatively meaningless.

Government take statistics and comparisons have inherent weaknesses as mentioned earlier. However, just as a reservoir engineer could never adequately describe the virtues of a reservoir without mentioning porosity, an analyst could not begin to characterize a fiscal system without the take statistic.

Figure 2.4 represents fairly well the universe of systems that existed during the late 1990s.[3] It includes the results of the recent feeding frenzy (January EPSA IV license round) in Libya. Some of the bars on the graph are wider than others because some countries have fixed terms (narrow bars) but many countries have either "bid" or "negotiated" terms and there is more variation and diversity found in the country's agreements. Also systems with "R factors" or "rate-of-return" (ROR) features can have a greater range of financial outcomes than more conventional systems. The universe of systems represented in Figure 2.4 was forged in an era when oil prices averaged a little over $18 per barrel (/bbl) and around 90% of the time ranged between $16 and $20/bbl. The natural question is "How do terms change with $60/bbl oil?" This is shown in Figure 2.5.

Figure 2.5 shows how systems change when oil prices increase three fold as they have in the past five years – from around $20/bbl to $60/bbl. Notice that with most systems the take only changes by a few points (2% to 3%). And, these systems are regressive (front-end-loaded) – so government take goes down. But some systems are progressive and government take goes up, typically by more than just a few points. The progressive systems are those with either an "R factor," a ROR feature or a price-cap formula. Many countries around the world right now wish they had structured their system to adjust their take upward.

Figure 2.6 caters to another natural question: "What kind of terms at $60/bbl would yield the same economic benefits as the terms found in Figure 2.5 at $20/bbl?" For an international oil company to achieve the same economic benefits or values (including *risked* values such as

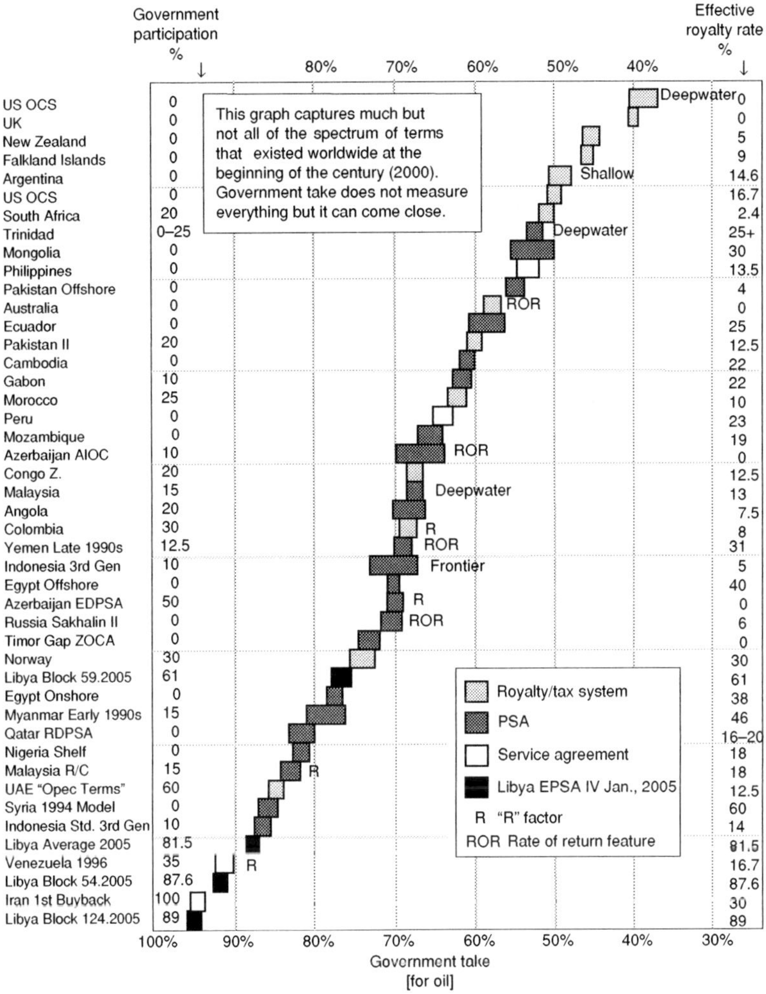

Source: Daniel Johnston & Co., Inc.

Figure 2.4 Government Participation, Effective Royalty Rate and Government Take (1)

"expected value") government take can be quite high. For example, from an international oil company point of view an average government take of 67% during the late 1990s at $20/bbl is roughly equivalent to a government take of 92% at $60/bbl. This is not to suggest that this is what countries should attempt to obtain. However, in the recent Libyan license rounds companies appear to have "bid" terms consistent with nearly

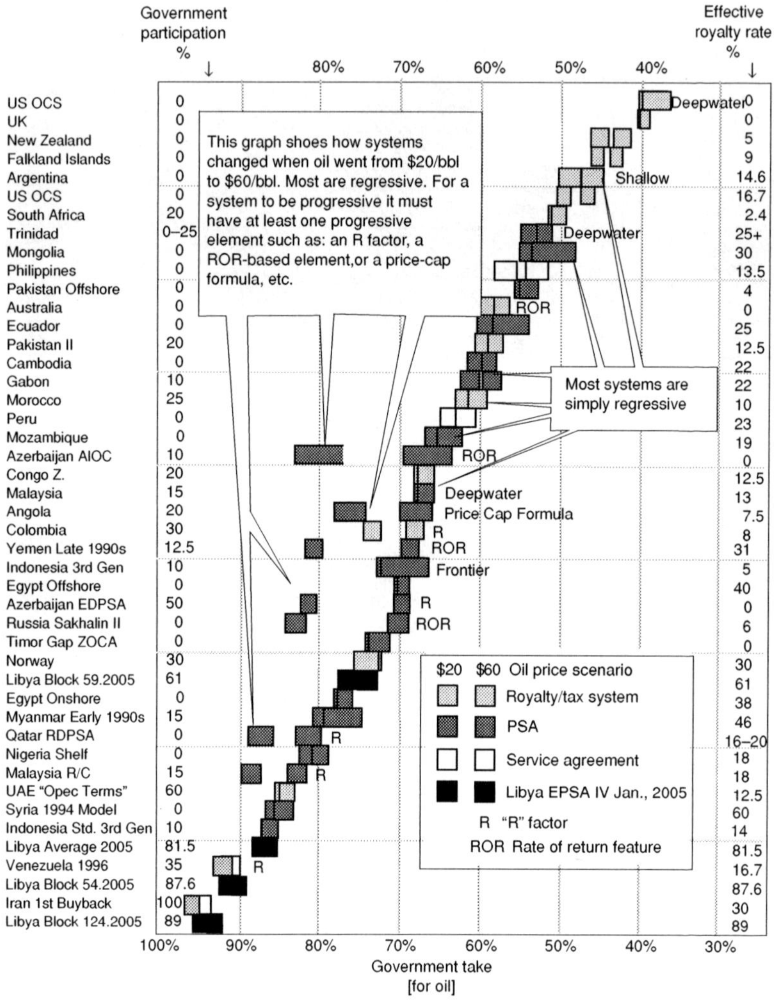

Source: Daniel Johnston & Co., Inc.

Figure 2.5 Government Participation, Effective Royalty Rate and Government Take (2)

$50/bbl expectations. This goes to the heart of the issue of allocation. By allocating licenses in a competitive bid round it was ultimately the IOCs who determined what the market could bear for the Libyan blocks. This took the burden of fiscal "design" off the NOC personnel and placed it upon the IOCs.

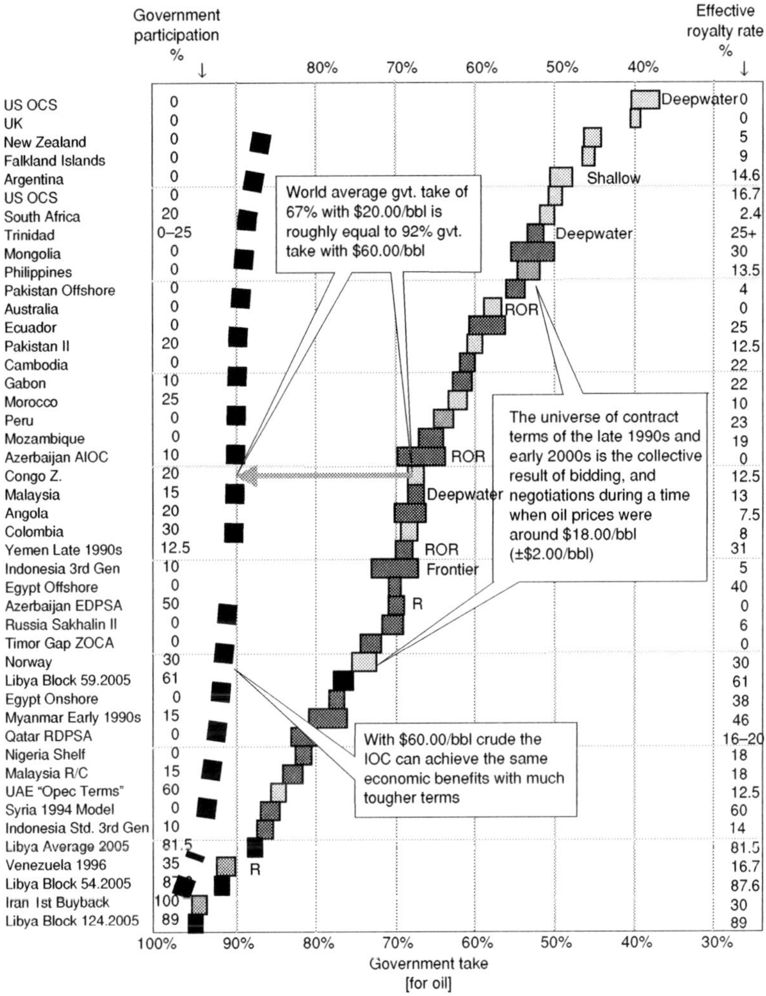

Source: Daniel Johnston & Co., Inc.

Figure 2.6 Government Participation, Effective Royalty Rate and Government Take (3)

Effective Royalty Rate (ERR)

The ERR is the minimum share of gross revenues a government will receive in any given accounting period for a field. It does not include the national oil company or oil minister's working interest share of production (not the way we calculate it), but it does give an idea of how front-end-loaded the system is, and gives contractors a feel for how quickly they will get their money back. This index was developed by Daniel Johnston in the mid 1990s and has become a standard metric in the industry. Wood Mackenzie sometimes refers to this as "minimum government take." It is an important index that adds dimension to the "take" statistics. The complement of ERR, "access to gross revenues" (AGR) provides an important international oil company perspective.

AGR is the maximum share of revenues a company or consortium can receive relative to their working interest in any given accounting period. It is limited by government royalties, and/or cost recovery limits and profit oil split (i.e. the ERR).

In a royalty/tax system with no cost recovery limit, the royalty is the only government guarantee. The ERR is the royalty rate. AGR is limited only by the royalty. In most royalty/tax systems in any given accounting period there is no limit to the amount of deductions a company may take and companies can be in a no-tax-paying position. But this can occur with a PSC as well.

Production sharing contracts with cost recovery limits guarantee the national oil company a share of profit oil because a certain percentage of production is always forced through the profit oil split. Thus both royalties and cost recovery limits guarantee the government a share of production or revenues regardless of whether or not true economic profits are generated.

The ERR/AGR calculations require a simple assumption – that expenditures and/or deductions in a given accounting period relative to gross revenues are unlimited. Therefore cost recovery is at its maximum (saturation) and deductions for tax calculation purposes yield zero taxable income. Situations like this can occur in the early stages of production, with marginal or sub-marginal fields, or at the end of the life of a field. The object of the exercise is to test the limits of the system. This provides the ERR/AGR indices.

Note: One key weakness of the ERR index is that it does not measure the effects of depreciation or amortization.

Also note: We do not include the effects of the guarantee provided by government participation if and where it exists.

Huge problems can arise if the effective royalty rate is not taken into consideration when designing a fiscal system. Depending on costs and production, contractors could go years where they are in a no-tax-paying position. This can cause cash flow problems for governments as well as lopsided misperceptions. We saw this in Ecuador in the mid 1990s because the ERR there under their service agreement was zero (0%).

Kazakhstan's Kashagan PSA has a government take of around 83% or more depending on a lot of things, but only a 2% ERR. The contract is said to be extremely complex and back-end-loaded. 83% is a very respectable take, but the government does not receive the bulk of it until the later years. It is estimated that in the first five to seven years of production, the government will only receive 2% of gross revenues. This is unlikely to be popular with the legislature, the press, or the citizens.

Effective royalty rate calculation

Figure 2.7 calculates the effective royalty rate of an Indonesian-type PSC. Again, one barrel of oil is used to represent revenues for a single accounting period. Typically this would be an early accounting period following production start-up when accumulated costs are high and production is relatively low.

10% royalty	Oil price $20/bbl
50% cost recovery limit	Costs assumed to be unlimited
60% government P/O share	
30% corporate income tax (CIT)	

Contractor share		Gross revenues $20.00		Government share
		Royalty 10%	→	$2.00
		$18.00		
$10.00	←	**Cost recovery** 50% limit		
		$8.00 **Profit oil (P/O)**		
$3.20	←	**Profit oil split** 40/60%	→	$4.80
($0.00)	←	**Tax rate** 30%	→	$0.00
$3.20				
$13.20		Division of gross revenues		$6.80
66%		Effective royalty rate		34%
$13.20/$20.00		Access to gross revenues		$6.80/$20.00

Figure 2.7 Indonesian-Type PSC – Effective Royalty Rate Calculation

In this example the contractor is in a no-tax-paying position, still the government receives a 10% royalty, and because of the cost recovery limit, the government is also guaranteed a percentage of the profit oil.

The 34% ERR in Figure 2.7 is high by world standards.

Figure 2.8 shows how contractors can be in a no-tax-paying position. The example shows an accounting period early in the development phase of a project where costs are high.

This really bothers some people. How it works:

Assumptions:

Discovery size (Recoverable)		**100 MMbbl**	
Exploration costs		**$50 MM**	(Expensed)
Development costs		**$400 MM**	
Depreciation		**25%/year**	
OPEX (operating expenses)		**$3.50/bbl**	
Production	**Year 4**	**4 MMbbl**	(Start-up)
	Year 5	**6 MMbbl**	(Ramp-up)
	Year 6	**11 MMbbl**	(Plateau)
Decline rate		**10%**	
Oil Price		**$20.00/bbl**	

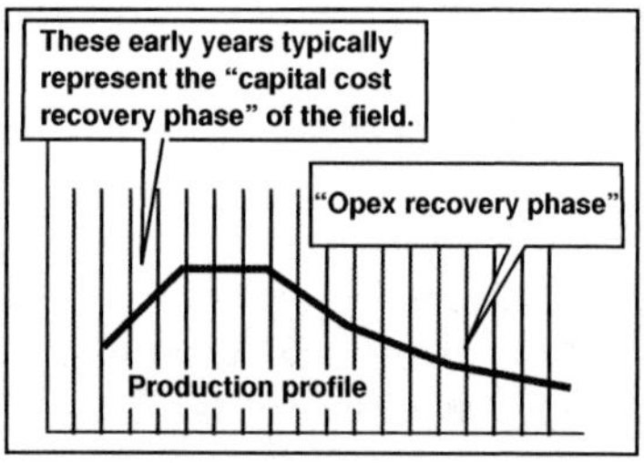

Tax Deductions for Year 4

Exploration costs	$50 MM	(Expensed)
Depreciation	$100 MM	(25%/year SLD)
OPEX	$14 MM	($3.5/bbl * 4 MMbbl)
Total	$164 MM	

Year 4 Revenue Distribution

$80.00	**Gross revenues ($MM)**
– 8.00	**10% royalty**
72.00	Net revenues
– 40.00	**Cost oil** (saturated – 50% limit)
32.00	Total profit oil
– 19.20	**Gvt. P/O share 60%**
12.80	Contractor profit oil
–0.00	**Income tax 30%**
12.80	Contractor P/O
+ 40.00	Contractor C/O
$52.80	66% of gross revenues

Year 4 Tax Calculation

$80.00	**Gross revenues ($MM)**
– 8.00	**10% royalty**
72.00	Net revenues
– 19.20	**Gvt. P/O share 60%**
52.80	Contractor revenues
– 52.80	Tax deductions (no limit)
0	Contractor tax base

TLCF into Year 5 = $164–52.8 MM
= $111.2 MM

Tax Deductions for Year 5

Tax loss C/F	$111.2 MM	($164 – $52.8 MM)
Depreciation	$100 MM	(25%/year SLD)
OPEX	$21 MM	($3.5/bbl * 6 MMbbl)
Total	$232.2 MM	

There won't be any taxes paid in year 5 either!

Source: Daniel Johnston & Co., Inc.

Figure 2.8 No Taxes???

This does not mean the government is not receiving revenue. In the case above, the government is receiving a royalty and 60% of the profit oil.

Regressive systems

A regressive system is one where government take goes down as profitability goes up. For a system to be regressive it must have at least one regressive fiscal element. Conversely, for a system to be progressive it must have at least one progressive element. Today, oil prices are more than double what they were when most of the existing fiscal systems were designed or negotiated. With the higher oil prices came higher profitability, but with most systems a lower government take. Governments are benefiting from the higher oil prices, total revenue does increase. But for most of them percentage share of net profit decreased, as seen earlier in Figure 2.5. This is simply a function of system design.

Many systems have sliding scales built into them to take advantage of the possibility of increased production, but few systems were designed to take advantage of the increased oil prices. The various elements of a fiscal system determine whether the system will be regressive or progressive (Table 2.4).

Table 2.4 Fiscal Systems: Elements and Effects

Element	Effect
Bonuses	Extremely regressive
Royalties	Very regressive
Taxes	Neutral
Government Participation	Neutral
"R" factors	Progressive
ROR systems	Progressive
Depletion allowances	Very progressive
Uplifts & investment Credits	Slightly progressive

Ownership

Title to hydrocarbons – One of the distinguishing characteristics of a petroleum fiscal system is where, when, and if ownership of the hydrocarbons is transferred to the IOC. From Table 2.5, we see that, with an R/T system, title transfers to the IOC at the wellhead. For a PSC, title transfers at the export point or *fiscalization* point. With service agreements (by definition) there is no title transfer.

Title to facilities – Title to facilities remains with the oil company under R/T systems, but under PSCs and service agreements it transfers to the NOC or government. There is some variation as to *when* title to facilities transfers to the NOC or government but usually this is upon commissioning of the facilities. In Nigeria title transfers as soon as equipment is landed in-country. Some countries will take title once the IOC has achieved payout.

Table 2.5 Fiscal Systems – Side-by-Side Comparison

	R/T systems	PSCs	SAs
Type of projects	*All types:* Exploration Development EOR	*All types:* Exploration Development EOR	All types but often non-exploration
Ownership of facilities	International oil company	Government NOC	Government NOC
Facilities title transfer	No transfer	"When landed" or upon commissioning	"When landed" or upon commissioning
IOC ownership of hydrocarbons (lifting entitlement)	Gross production less royalty oil	Cost oil + profit oil	None
Hydrocarbon title transfer	At the wellhead	Delivery point Fiscalization point or export point	None
Financial obligation	Contractor 100%	Contractor 100%	Contractor 100%
Government participation	Yes Not common	Yes Common	Yes Very common
Cost recovery limit	No	Usually	Sometimes
Government control	Low typically	High	High
IOC lifting entitlement	Typically around 90%	Usually from 50–60%	None (by definition)
IOC control	High	Low to Moderate	Low
% of systems (legal point of view)	44%	48%	8%
% of systems (philosophical point of view)	44%	56% – from a philosophical perspective all of these are **Service agreements**	

Lifting entitlement and reserves reporting

As a general rule oil companies will "book" barrels according to (1) their working interest and (2) their lifting entitlement. There are two main exceptions to this rule: (1) even though by definition under a service agreement there is no entitlement, companies do book barrels, (2) with systems where taxes are "in lieu" companies "gross-up" their actual entitlement and book the barrels they would have been entitled to lift had they paid taxes directly in cash. This is common with "Egyptian-type PSCs." R/T systems would be much preferred by an IOC wanting to "book barrels" because they can typically book about twice as many barrels as they would with a PSC. Because a company's entitlement with a PSC is based on its share of cost oil and profit oil, when oil prices went from $20/bbl to $60/bbl the typical entitlement under a PSC went down by around 15%. This is because with higher prices it does not take as much cost oil to recover costs and thus entitlement goes down. This is a phenomenon we do not see with R/T systems.

Few international oil company shareholders or government officials understand or appreciate the weaknesses, inconsistencies and complications associated with "booking barrels" or "modern reserves disclosure" requirements, which have not changed appreciably since the late 1970s.

Allocation

Oil companies will suffer just about anything for highly prospective acreage or projects. Sealed bid license rounds (auctions) can be very beneficial for a government with highly sought-after acreage or projects. However, even countries like Nigeria and Kazakhstan have some acreage and some projects that are not quite as exciting as others. When it comes to attracting IOC investment, allocation becomes much more important with less-than exciting prospects. When it comes to this issue one of the most difficult things for IOCs to contemplate is a direct "heads-on" competitive sealed-bid license round for non-spectacular acreage or projects.

Allocation and license promotion

A theme developed a few years ago stated that acreage has begun to take on more of the characteristics of a commodity. There is over three times as much acreage available today as there was 25 years ago. In the past two decades the Soviet Union has become the "former" Soviet Union (FSU) and much of Africa and the Eastern-bloc countries have opened up. Furthermore, with more aggressive and specific relinquishment provisions in contracts the market for acreage or projects is more dynamic and robust. It is appropriate to think of acreage as a commodity and a global commodity at that. Countries are competing with more than just their neighbors for capital and technology.

Governments have no control over what nature has chosen to bequeath to them. However, they do have some control over the industry's perception regarding those gifts. This becomes especially important if the government believes the industry has an unfair view of its country's potential.

The means by which governments determine how to award licenses are extremely varied. Some governments (approximately 30 to 40 each year) have official "block offerings" or "license rounds" where blocks are awarded on the basis of competitive bids. Other countries will negotiate exploration rights one-on-one with companies. While companies typically prefer negotiated deals, these situations can be just as competitive as an official tender. It depends on the prospectivity of a block or area. With less-than-exciting acreage a government may have no choice – negotiated deals may be the only option.

While allocation strategy is not as important as, for example, the commercial terms, it can add an important dynamic for governments competing for capital and technology. When Venezuela launched its exploration round in 1996, it put ten blocks up for bid. However, for all practical purposes Venezuela had ten separate license rounds, block-by-block. On Monday morning January 22, 1996, bids were opened for the first block only (the La Cieba block). These licenses were awarded on the basis of a single-parameter bid – a profits-based tax known as the "PEG." Companies were to bid from zero to a maximum of 50%. Royalty and other fiscal elements were "fixed" (i.e. neither biddable nor negotiable). Ties were to be broken by a subsequent bonus bid round to follow the opening of the PEG bids within a couple of hours. On the first block, La Cieba, 11 companies bid and nine tied with a full 50% PEG bid. The tie was broken with a bonus of $103,999,999 from the Mobil/Veba/Nippon consortium. That afternoon the next license (Paria West) was awarded to Conoco under the same rules. This kind of approach magnified the already intense competition by awarding licenses one at a time. With each "round" the pool of bidders would potentially be reduced by perhaps only one group if any. This approach greatly reduced the chance that less-prospective blocks would receive no bid. There were, however, two blocks that did not receive a bid. The resulting government "takes" were around 92%.

At the other end of the spectrum, in the United States Gulf of Mexico, licenses are awarded solely on the basis of a bonus bid. In fact, few countries worldwide extract such a large portion of rent through bonuses.

Things to consider when it comes to allocation:

- Fixed terms or negotiated terms or bid terms
- Block size and configurations
- Timing/duration/relinquishment provisions
- Work programs
- Ringfencing
- Proactive marketing – Road shows?
- Data availability – Data fees – Data rooms?
- Permitting requirements
- Signal theory

Table 2.6 Different Situations – Different Considerations

	Enhanced oil recovery	Development projects	Exploration acreage	Frontier acreage
Degree of risk	Med. to High	Low	High	Highest
Block size **Acres** **(km^2)**	Field 4,000 or so (16)	Smaller 3,000–5,000 (12–20)	Large 1–2 MM+ (8,000)	Very large 3–4 MM+ (16,000)
Work program(s)	1) Feasibility study 2) Pilot program 3) Development	1) Appraisal 2) Development	Exploration program	Exploration program
Focus of negotiations/ analysis	IRR	IRR	Take	Take
Most common allocation strategy	Negotiated deals	Negotiated deals	Competitive bidding and other means	Competitive bidding and other means

Fixed terms or negotiated terms or bid terms

This issue is of huge concern to many governments. Many governments, through their national oil company or oil ministry, will "fix" the key fiscal terms (such as royalties, profit oil share and taxes). There is no bidding or negotiation of "fixed" terms.

There is considerable pressure these days from the World Bank, the International Monetary Fund and even the UK's Extractive Industry Transparency Initiative (EITI) for oil companies and governments to be more open and disclose more information, to be more transparent and "publish what they pay." With these initiatives there is a strong push for governments to be more transparent and allocate acreage on the basis of public auctions similar to the highly publicized EPSA IV rounds in Libya in 2005.

The problem is that unless the acreage is particularly interesting, the industry has been relatively unwilling to face the kind of magnified "head-on" competition that a "sealed bid" type license round (like Libya) provokes. I believe that in many instances it is naïve and unrealistic to expect all governments to allocate acreage and projects on the basis of sealed bids.

Allocating licenses through "negotiated deals" can be efficient too. Government officials (energy ministry or NOC) become aware of what the market can bear as they entertain various proposals and offers. Likewise the lack of interest provides information too. For a NOC official there is nothing worse than a "failed license round."

Elements that become part of a contract or a system are usually either: negotiated, statutory or a bid item.

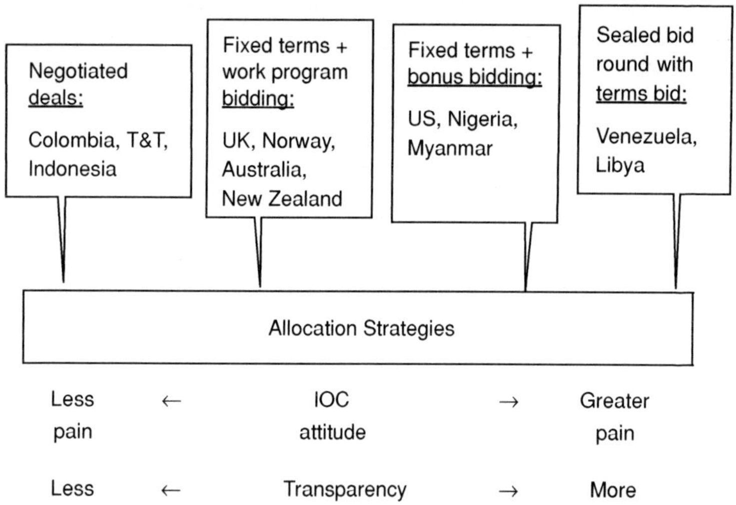

Source: Daniel Johnston & Co., Inc.

Figure 2.9 License/Acreage Allocation/Auction Spectrum

Government participation

Government participation is a controversial element of the government take statistic. Government participation typically is the result of a government option (through the NOC) to take up a working interest in the event of a commercial discovery. Over half of the countries worldwide have this option. Once the government exercises the option it then "pays its way" for development and operating costs from the commerciality point forward just like any other working interest partner. Thus the government is "carried" through the exploration and appraisal phase. This is also called a "back-in." Around half of the NOCs will reimburse their share of "past costs" after they "back-in"; the other half do not. "Past costs" are defined as those costs incurred by the IOC after the "effective date" of the contract up to the "commerciality date" when the NOC "backs-in." The main source of the controversy is the issue of whether or not it should be included in the take calculation – that is, "Is it truly a rent extraction mechanism?"

Government participation analysis controversy

One of the more controversial aspects of fiscal system analysis is the treatment of government participation or the back-in option. Certain analysts, including some at the World Bank, believe it is not appropriate to view this element of a system as a rent extraction mechanism. The argument goes like this:

> Government take as a result of equity participation by government is really a government equity return, directly paid for by government, rather than a form of government take. Hence, comparing government take statistics by excluding government equity participation is probably a more accurate representation of levels of take.[4]

However, this approach contradicts some basic economic laws. And, it is easy to check by asking a simple question: *Does typical government participation (i.e. the "back-in") cause an international oil company or foreign investor any financial pain?*

The answer is a resounding "Yes." And the pain is multidimensional. First of all the value of a discovery to an explorer will be reduced by almost exactly the amount of the "carry" and secondly, the companies will not be able to "book" as many barrels.

Certainly there is a difference between say a 50% profits-based tax and a government back-in option of 50% – both of which will guarantee the government an added 50% share of profits. An IOC would happily avoid both if it could. From a project cash flow point of view, companies will certainly prefer 50% government participation to a 50% tax because at least with participation, after the NOC backs-in, it "pays its way." Just *how* different the financial impact is between a 50% tax and a 50% back-in depends on profitability and timing. As profitability increases the back-in or participation element takes on more of the characteristics of a pure tax or a royalty depending on the point at which the government takes its share of production. While it is conceptually a bit abstract, as costs relative to gross revenues approach zero (the ultimate in profitability) the back-in begins to take on *all* of the characteristics of a tax. Thus, the less profitable a venture is, the less painful the government participation element is. Either way though, both taxes and/or participation options cause the contractor financial pain to various degrees.

Comparing two fiscal systems on the basis of government take alone is not a perfect comparison if one system has participation and the other does not. This highlights one of the key weaknesses of typical undiscounted government take statistics. However, to simply ignore the participation element would be a greater misrepresentation. When comparing fiscal terms for exploration rights it is not appropriate to exclude or ignore the participation element as the argument above suggests.

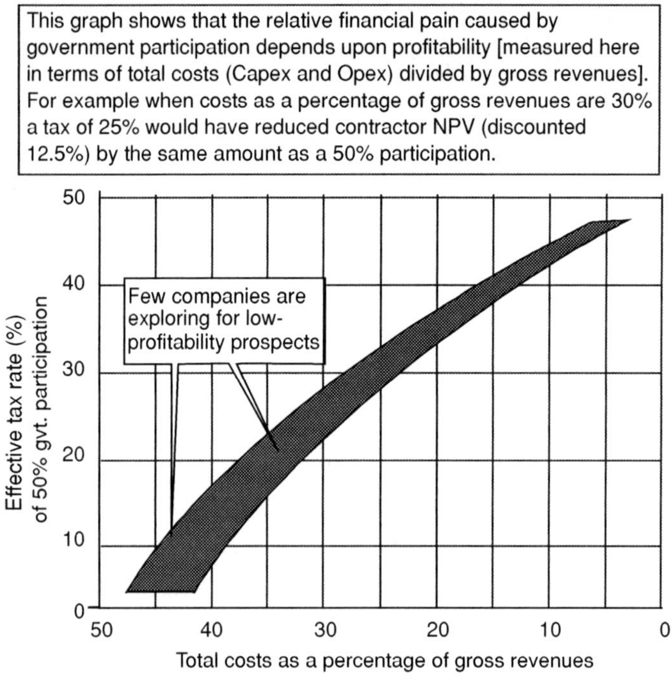

Source: Daniel Johnston & Co., Inc.

Figure 2.10 Government Participation vs Effective Tax Rate

Many systems provide an option for the national oil company to participate in development projects. Under most government participation arrangements, the contractor bears the cost and risk of exploration and if there is a discovery the government backs-in for a percentage. In other words the government is carried through exploration. This is fairly common and automatically assumed whenever some percentage of government participation is quoted.

Technically the government, through the NOC, is carried through by a well or two from the actual discovery well. The contract clause that deals with the requirement for delineation/appraisal wells following a discovery is referred to as the "commerciality clause." The government agent, usually the NOC, must decide whether to exercise their right to "back-in" after the discovery has been appraised – the "commerciality point."

Over half of the countries with any respectable hydrocarbon potential have the option to back-in at the point of commerciality.

The key aspects of government participation are:

• What percentage participation?	(Most range from 10% to 50%) (Average is around 30%)
• When does the government back- in?	(Usually at commerciality)
• How much participation in management?	(Large range)
• What costs will the government bear?	(Usually their pro rata share of costs)
• How does government fund its share of costs?	(Often out of up to a certain % of the government's share of production)
• Do governments reimburse "past costs"?	(Half do – half do not)

The financial effect of a government partner is similar to that of any working interest partner with a few large exceptions. First, the government is usually carried through the exploration phase and may or may not reimburse the contractor for past exploration costs. Second, the government contribution to capital and operating costs is often paid out of production. Finally, the government is seldom a silent partner.

In Colombia the government has the right to take up to 50% of working interest and will reimburse the contractor up to 50% of any successful exploratory wells. In China, the government participation is 51%. This usually defines the upper limit of direct government working interest involvement.

Contractors prefer no government participation. This is not totally selfish, but stems from a desire for efficiency as well as economy. Joint operations of any sort, especially between diverse cultures, can have a negative impact on operational efficiency.

How these different arrangements function in different systems

Royalty/tax systems

Prior to the late 1960s R/T systems, or concessionary systems, for all practical purposes were the only arrangements available. Typically, the key points of such a system were:

- Oil companies contracted for the right to explore for hydrocarbons
- If a discovery was deemed to be commercial, the international oil company had the right to develop and produce the hydrocarbons
- When hydrocarbons were produced the international oil company would take title to its share – gross production less the royalty – at the wellhead
- Entitlement meant gross production less royalty. If the royalty was 10% the IOC could "lift" 90% of production. If the royalty was paid in cash, then the IOC could "lift" 100% of production
- Exploration and production equipment was owned by the IOC
- IOCs paid taxes on profits from the sale of the oil.

Sample calculation The following example (Figure 2.11) demonstrates the arithmetic performed to calculate contractor and government take, and entitlement under an R/T system. Even though this analysis represents "full cycle" figures the hierarchy of arithmetic that would be expected in any given accounting period is honored. In this particular case $20/bbl is assumed to represent average gross revenue per barrel over the life of the field (full cycle).

In this example R/T system, we calculate government take over the full cycle of the project, which includes exploration and early development through to field decline and abandonment. We use one barrel of oil, or $20, to represent average full cycle revenues (per barrel) and show how that barrel of oil is divided between the government and the contractor.

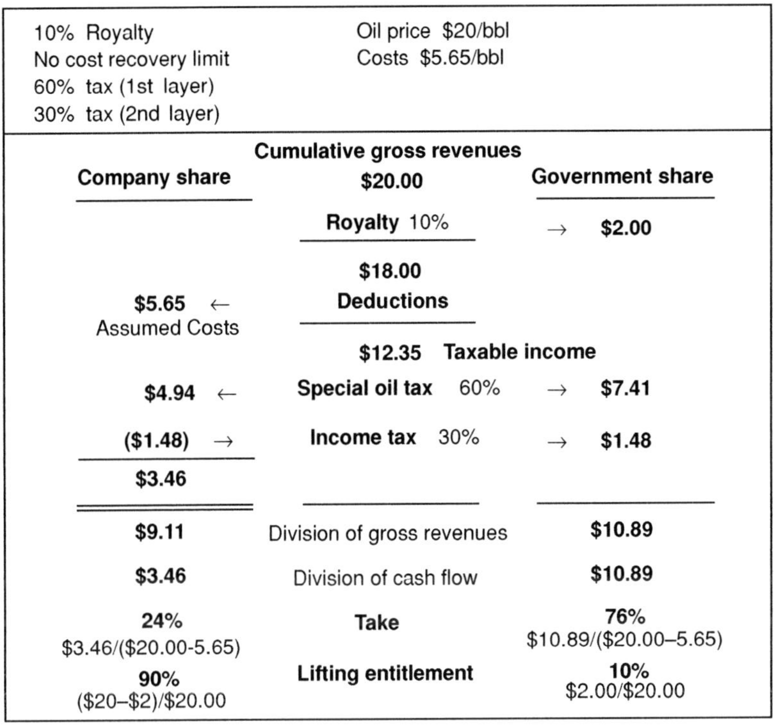

Figure 2.11 Royalty/Tax System Flow Diagram – One Barrel of Oil (Full Cycle)

The government gets a 10% royalty = $2. Assumed costs ($5.65) are deducted from the $18 left after the royalty is taken, leaving a taxable income of $12.35. Two layers of taxes are levied against the taxable income: first a 60% tax on the $12.35 gives the government $7.41, leaving $4.94. The second layer of tax, 30%, is levied against the $4.94, giving the government an additional $1.48 and leaving the contractor with $3.46.

Take statistics are a function of cash flow (gross revenue – costs). And in this particular example government take = government cash flow/total cash flow or $10.89/($20.00–$5.65) or 76%.

Production sharing contracts

The concept of production sharing is ancient and widespread. Farmers in the USA have been familiar with the concept for decades. As far as the oil industry is concerned (and according to a Permina brochure dated 2000, the concept was conceived in Venezuela in the mid 1960s), the first modern PSC was signed in 1966 between IIAPCO and Permina Indonesia's national oil company at the time. The characteristic features of this pioneering agreement included:

- Title to the hydrocarbons remained with the state (Indonesia)
- Permina maintained management control
- [Note: Putting management control in the hands of Permina is what really distinguished the PSC from earlier Indonesian arrangements, known as the Contracts of Work (COWs)]
- Contractor submitted work programs and budgets for approval
- Profit oil split (P/O) 65%/35% in favor of Permina
- Contractor bore the risk
- Cost Recovery (C/R) limit was 40%
- Taxes were paid in lieu (i.e. taxes paid for and on behalf of the IOC by Permina)
- Purchased equipment became the property of Permina
- Company entitlement = cost oil + profit oil.

Sample calculation The following example (Figure 2.12) demonstrates the arithmetic performed to calculate contractor and government take, and entitlement. In this case, like the example R/T calculation above, we use the revenue from one barrel of oil – $20 – to represent average (per barrel) gross revenue over the life of the field (full cycle).

This example is mathematically identical to the previous R/T system example – with the obvious exception of lifting entitlement. The terminology however is different. In the R/T system we use the term deductions, whereas with PSCs the term "cost recovery" is used, and instead of a 60% tax, there is a 60/40 profit oil split in favor of the government. Still, the arithmetic is the same and government and contractor take calculations are identical to those in the R/T system. This illustrates that from a mathematical/mechanical point of view the differences between R/T systems and PSCs are by far outweighed by the similarities.

From a mechanical point of view the C/R limit is the only difference between R/Ts and PSCs.

Entitlement is another matter. In this example, otherwise identical to the previous R/T system over the full cycle, contractor entitlement (53%) is about half that of the R/T system, where the contractor was entitled to lift 90%.

Service agreements

Service contracts or service agreements generally use a simple formula: the contractor is paid a cash fee for performing the service of producing mineral resources. All production belongs to the state.

10% royalty 50% cost recovery limit 60% government P/O share 30% corporate income tax (CIT)	Oil price $20/bbl Costs $5.65/bbl

Contractor share	Cumulative gross revenues $20.00	Government share
	Royalty 10%	**$2.00**
	$18.00	
$5.65 Assumed costs	**Cost recovery** 50% Limit	
	$12.35 Profit oil	
$4.94	**Profit oil split** 40/60%	**$7.41**
($1.48)	**Tax rate** 30%	**$1.48**
$3.46		
$9.11	Division of gross revenues	**$10.89**
$3.46	Division of cash flow	**$10.89**
24% $3.46/($20.00-5.65)	**Take**	**76%** $10.89/($20.00-5.65)
53% ($5.65+4.94)/$20.00	**Lifting entitlement**	**47%** ($2.00+7.41)/$20.00

Figure 2.12 Typical PSC – Flow Diagram – One Barrel of Oil (Full Cycle)

The contractor is usually responsible for providing all capital associated with exploration and development (just like as with R/T systems and PSCs). In return, if exploration efforts are successful, the contractor recovers costs through the sale of oil or gas plus a fee, which is often taxable. These agreements can be quite similar to PSCs or R/T systems except for the issue of entitlement. As mentioned earlier, other than the issue of entitlement the Venezuelan terms from the 1996 round look and sound just like an R/T system – they have a royalty and taxes. But the Philippine SA uses the terminology and structure of a PSC with a cost recovery limit and profit oil split.

Following are examples of various SA fee structures.

Fixed fee – $/bbl A $/bbl fee-based formula is used in joint ventures in Nigeria, a few contracts in Abu Dhabi, and is also part of Kuwait's proposed Operating Service Agreement (OSA). A simplified example is summarized as follows: first, the IOC conducts operations in much the same way as it would in virtually any fiscal system. For performing these services (in this example) the IOC is able to recover its costs (assumed to average $4/bbl) out of revenues and is also paid a $2/bbl fee for conducting operations. Example 1 shows how this simple arrangement looks at $20/bbl and $60/bbl oil prices.

Example 1:

A	$20.00	$60.00	Gross revenues ($/bbl)
B	-2.00	-2.00	Fee $2/bbl
C	18.00	58.00	Net revenue
D	-4.00	-4.00	Assumed costs
E	14.00	54.00	Government profit (cash flow)
F	2.00	2.00	Company cash flow
	87.5%	96.4%	Government take [F/(A–D)]
	12.5%	3.6%	Company take [B/(A–D)]

Notice with this structure the system is "progressive": as oil prices go up (or as profitability goes up) government take also goes up.

Fixed fee – % of costs – uplift Another type of fee-based approach – like that found in Iran under the "buy-backs" and proposed in Iraq under what is called a "squeeze PSC," which sounds a lot like a "buy-back" – provides the IOC with a means of recovering costs plus a fixed fee that is a function of the anticipated costs. Example 2 assumes the IOC will be reimbursed for costs of $4/bbl plus an uplift of say 50% (of those costs). This is a simple example but it should serve our purposes. The IOC would conduct operations in much the same way as with other petroleum operations. The example here shows how this arrangement would look with oil prices of $20/bbl and $60/bbl.

Example 2:

A	$20.00	$60.00	Gross revenues ($/bbl)
B	-4.00	-4.00	IOC cost recovery (reimbursement)
C	-2.00	-2.00	IOC fee 50% of costs (remuneration)
D	14.00	54.00	Government profit (cash flow)
C	2.00	2.00	Company cash flow
	87.5%	96.4%	Government take [D/(A–B)]
	12.5%	3.6%	Company take [C/(A–B)]

Notice this system is also "progressive": as oil prices go up (or as profitability goes up) government take goes up.

Variable fee – percentage of gross revenues Another type of fee-based approach (very rare) provides the IOC with a direct share of revenues from which, hopefully, it would be able to recover its costs and make a profit. This type of arrangement in its classic form is referred to as the "Peruvian model." Another variation is the Filipino Participation Incentive Allowance (FPIA), which allows the contractor group a 7.5% "incentive" if there is sufficient "Filipino participation." This 7.5% allowance is based on gross revenues. A simple example here assumes the IOC will receive 25% of gross revenues. The IOC conducts operations in much the same way as it would under almost all petroleum systems. Example 3 shows how this simple arrangement looks at $20/bbl and $60/bbl oil prices.

Example 3:

A	$20.00	$60.00	Gross revenues ($/bbl)
B	-5.00	-15.00	IOC fee 25% of gross revenues
C	15.00	45.00	Government profit (cash flow)
D	-4.00	-4.00	Assumed costs
F	1.00	11.00	Company cash flow (B–D)
	93.75%	80.4%	Government take [C/(A–D)]
	6.25%	19.6%	Company take [F/(A–D)]

Notice that with this structure the system is "regressive." As oil price or profitability goes up, government take goes down. This is because while the IOC is guaranteed 25% of gross revenues (almost like a negative royalty) the government is guaranteed 75%, like a large royalty. Royalties are notorious for being regressive, especially large royalties.

The key differences between SAs and PSCs There are numerous sources or references that make little distinction between PSCs and SAs other than differences regarding the transfer of title to hydrocarbons (discussed earlier). The one most common distinction between the three main families of fiscal systems is the issue of title transfer of hydrocarbons from the state to the IOC.

With the classical legal definition of a SA the IOC does not take title to hydrocarbons as it does with a PSC. One implication is that the IOC may likely "book barrels" in different quantities under a SA than it would with a PSC.

To summarize:

- With royalty/tax systems title transfers at the wellhead
  The IOC takes title to gross production less royalty oil

- With PSCs title transfers at the export point or "fiscalization point"
  The IOC takes title to cost oil and profit oil
- With service agreements, title does not transfer.

Physical asset ownership and ownership transfer With both SAs and PSCs the government (usually through the national oil company) takes title to all production facilities, pipelines and other associated facilities. The important legal implication regarding facilities ownership is that the obligation for site restoration, abandonment and clean-up is held by the *owner* in the absence of clear and well-crafted abandonment provisions.

Usually under SAs and PSCs title to production facilities and pipelines and so forth transfers to the NOC (or the equivalent) when the facilities are commissioned. In some countries title transfer occurs sooner. For example, in Nigeria title to facilities transfers to the Nigerian National Oil Corporation (NNPC) when the equipment is landed in-country. Some countries will wait until the facilities have achieved "payout" at which point title transfers to the NOC.

From a financial point of view as far as normal production operations are concerned, there is little difference to the IOC whether they or the government own the facilities. The significant difference involves the abandonment/site-restoration liability.

What are the key concerns of most oil companies?

Oil companies want to explore in regions where there is a reasonable chance of finding oil, and sometimes gas. They want to deal with stable governments, and prefer contract "terms" that will provide a potential return-on-investment that is commensurate with the associated risks.

Oil companies are also interested in – or rather psychotic about – booking barrels. In the eyes of Wall Street, oil companies are measured by their ability to replace production and by finding costs. If they can book more barrels their "reserve-replacement ratio" benefits and their finding costs go down. And they are measured on finding and lifting costs and reserve replacement. This can be confusing and frustrating since the ability to book barrels and the amount of barrels a company can book strongly depends on the type of system and various other illogical elements.

What oil companies want:

- Excellent or good geology
- Excellent terms
  - Company take
  - Getting their money back as soon as possible (effective royalty rate)
- Stable contract terms
- Low levels of transaction costs with government

What oil companies get:

- Fair to good geology
- Tough terms
- Unstable situations
- Grief

Entitlement

For most companies the reserves they book will correspond to the reserves they are entitled to "lift." However, this is not a critical requirement. So naturally, there are some interesting exceptions:

Royalty/tax systems – Entitlement equals gross production less royalty oil
However, many governments take their royalty "in cash" instead of "in kind." In this case many companies are booking those barrels as well.

PSCs – Entitlement equals profit oil + cost oil
However, with the Egyptian-type systems where taxes are "in lieu" the companies are calculating what their profit oil share would have been (dividing their share by 1 minus the tax rate) and booking these "imputed barrels."

Also some companies are booking gas or oil consumed on-site for fuel for operations.

There are big differences between R/T systems and PSCs when it comes to entitlement. PSC entitlements typically go up with falling oil prices and down with increasing oil prices. This is not a problem with R/T systems.

Service agreements – By definition there is no entitlement
Nevertheless, companies are booking barrels under service agreements.

Savings index

The savings index is a measure (from an undiscounted point of view) of how much a company gets to keep if it saves one dollar. Because of the great concern on the part of both governments and companies about reducing costs, this statistic is included to quantify somewhat the incentive companies have to keep costs down. Only the profits-based fiscal elements influence this statistic. Royalties (based on production not profits) have no influence. The example R/T system has two profits-based mechanisms: a 60% special petroleum tax and a 30% income tax. Therefore, if the company saves one dollar then there will be an added dollar of taxable income. The government gets 60% of that. The company therefore has 40¢ on the dollar saved prior to implementation of the income tax. With a 30% income tax the company only gets to keep 70% of the 40¢. The savings index then is 28¢ on the dollar (saved), or 28% (see Example 3).

> *Example 3:*
> Here is what happens to a "dollar saved."
> A dollar saved means an extra dollar of profit oil.

$1.00	Profit oil
– 60	Government share
40	Taxable income
– 12	30% Income tax
28¢	Company cash flow

This index does not take into account present value discounting. The present value effect is interesting and it often magnifies the IOC's incentive to keep costs down.

Block size, relinquishment and ringfencing

These are elements that determine to a certain extent how much exposure to exploration risk the government will experience. Most governments go to a lot of effort to distance themselves as much as possible from exploration risk.

Block size and configurations

The choice of block size and configuration is an important consideration. For example, with larger blocks there is the likelihood of a greater accumulation of exploration sunk costs prior to discovery. These costs are typically recoverable and/or tax deductible and because of this with larger accumulations of sunk costs governments will receive fewer taxes, etc. With smaller blocks governments can minimize or mitigate their exposure. However, in frontier regions companies need larger tracts of acreage. Another challenge is to configure the blocks or licenses in order to provide interesting tracts instead of having just a few highly prospective blocks and others that will attract little interest.

Block sizes range from small to huge as the examples in Figure 2.13 illustrate. Typically, block sizes will be smaller in proven geological provinces and much larger in frontier regions. The larger regions can require considerable exploration expense. However, the IOC may be able to recover dry hole and other exploration costs in one part of a block against a production in another part of the block.

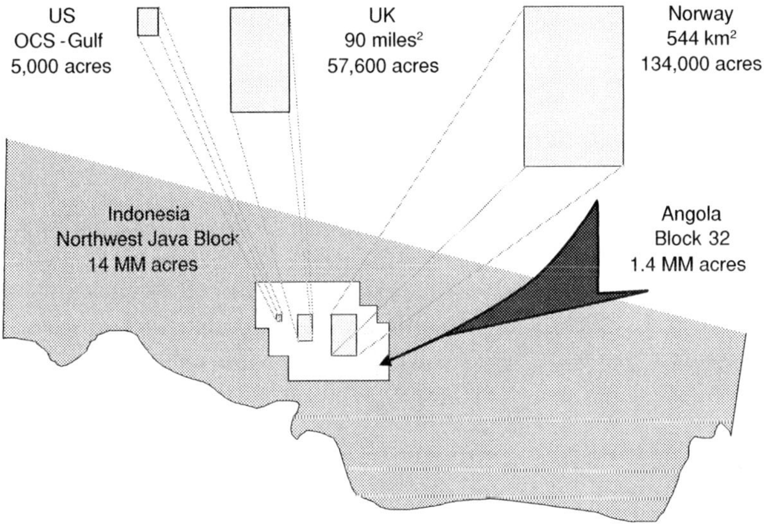

Figure 2.13 Block Sizes

Timing/duration/relinquishment provisions

These elements are also important. Relinquishment options are diverse and a full spectrum of methods is employed, ranging from almost no relinquishment (in the ordinary sense) to very aggressive relinquishment requirements such as we see in the Middle East. For example, in some of these countries only a discovery will be retained and all other acreage will be surrendered at the end of the final exploration stage. On the other hand, in Indonesia for many years oil companies could keep more than just development areas (discoveries) at the end of the final official stage of exploration. This meant that if a company made an economic discovery it could enjoy the opportunity to continue exploration in its *remaining* acreage while it pursued development of its discovery.

Ringfencing

Ringfencing is a term used to describe situations where governments will not allow companies to "consolidate" their operations from one license area to another. It means that each license (typically) is treated as a separate cost center for cost recovery and tax calculation purposes. A number of countries will automatically ringfence any discovery as soon as it is made, thus disallowing deductions for exploration activity outside the initial discovery area. This kind of treatment is becoming more and more common.

Ringfencing can protect a government from what might otherwise be a marginal or sub-marginal discovery, by limiting the costs that can be cost recovered and/or deducted against revenues generated by the discovery. However, it can be a negative incentive to the exploration companies.

Petroleum fiscal/contractual statistics

The statistics in Tables 2.7 and 2.8 are from a 2001 database.[5] Therefore they do not take into consideration the recent oil price increases. But, keep in mind that if most fiscal systems in the world are regressive, government take on average will go down slightly (see Figure 2.6), although the actual revenue governments receive will still go up.

Table 2.7 World Average Fiscal Terms

	PSCs and SAs	World average	R/T systems
Number of systems	72	136	64
Government take	70%	65%	59%
Government participation	36 countries	65 countries	29 countries
Royalty rate	5%	7%	8%
Effective royalty rate	23%	17%	8%
Ringfenced systems	75%	55%	30%
Lifting entitlement	63%	77%	92%
Savings index	39%	47%	56%
Cost recovery limit	65%	N/A	N/A
Systems with ROR or "R" factors	17%	21%	25%

Source: D. Johnston, *International Petroleum Fiscal Systems*, PennWell Books, 2001

Table 2.8 World Average Terms for Countries in the Top 20th Percentile (Prospectivity)

	PSCs	World average	R/T systems
Number of systems	19	25	6
Government take	78%	79%	80%
Government participation	12 countries	17 countries	5 countries
Royalty rate	5%	6.8%	11%
Effective royalty rate	29%	24.5%	11%
Ringfenced systems	90%	76%	33%
Lifting entitlement	55%	63%	89%
Savings index	30%	31%	37%
Cost recovery limit	62%	N/A	N/A
Systems with ROR or "R" factors	26%	24%	16%

Source: As Table 2.8

Recent developments around the planet

This section looks at recent changes that affect our industry globally: changes to the rules governing regressive systems; developments in Russia; acreage grand openings since the early 1990s; Bolivia's gas war of 2003; Iranian "buy-backs"; the emergence of gas; Mexico's multiple service contracts; changes in booking barrels in Venezuela; Libya's EPSA IV; the Sao Tome JDZ license round in Nigeria; and APPA harmonizing terms.

Changes to the rules governing regressive systems

$60 per barrel oil has pointed up problems with regressive systems, causing countries and regions with regressive and/or neutral fiscal systems to change their terms:

- **Alaska** is looking to eliminate the Economic Limit Factor (ELF)
- **UK** has raised taxes from 40% to 50%
- **Venezuela** has eliminated R/T credit
- **Bolivia** has radically changed the rules
- See also **Kazakhstan**, and **Trinidad and Tobago.**

Russia

During the 1990s the Russians considered two types of production sharing agreement: typical Indonesian-type systems like the Kharyaga and two Sakhalin PSAs, and what they called "direct sharing" systems like the old "Peruvian model" where gross production would be divided.

They found that a number of stumbling blocks emerged:

The issue of stability

IOCs wanted more contract stability than the Russians were willing to provide.

The unsustainability of PSAs

Mikhail Khodorkovsky and others convinced many people that PSAs were "unsustainable" because (they argued) oil companies under PSAs could typically recover all their costs before the state received anything. (By the way, this is not a fair generalization – it is an ERR issue and PSAs typically have higher ERRs.) Furthermore, there is no particular reason why a PSA had to have a low (or zero) ERR.

The "what club should we belong to?" argument

R/T system proponents argued that PSAs were not being used by other superpowers – so why should Russia be expected to use them? They pointed out that the systems of choice by other superpowers were R/T systems. And, they argued, what was good enough for the superpowers was good enough for Russia.

PSAs and Russian petroleum laws

Another important consideration forwarded by industry experts points up the problems PSAs had with existing Russian petroleum legislation. R/T systems simply fit better. PSAs were thought to be the system of choice for larger, problematic development projects. But, introducing PSAs across the board would require massive change to Russian petroleum legislation.

> As oil prices are high, tax revenue will be high. If in 2006 oil prices fall, taxes should be increased.
>
> Alexei Saratyugin, Director of Fiscal Policy of Ministry of Finance[6]

Acreage grand openings since the early 1990s

Competition is growing for the world's limited exploration capacity. Brazil, Venezuela, Iran, Kuwait, Russia, Kazakhstan, Azerbaijan, Turkmenistan and Libya have all opened acreage to exploration, with the result that there is roughly twice as much acreage available today.

Bolivia's gas war of 2003

The breakdown in Bolivia is an example of virtually nothing going right in the development of a country's resources. Bolivia is gas prone, landlocked, and poverty-stricken. Its poor, in the north, expected little benefit from gas development in the south, and protested by shutting down the roads leading to the capital, La Paz, in October 2003, the First Gas War. Since then the country has had three presidents and at least six oil ministers.

The new president, Evo Morales, wants to export gas and realizes that this will require help from outside companies. But, Morales promised to "nationalize" Bolivia's gas resources. One option is to develop a careful distinction between "nationalization" and "expropriation," and allow companies to export gas, but not book reserves.

Iranian "buy-backs"

The Iranian buy-backs have many of the characteristics of a pure service agreement – and are often referred to as short-term Risk Service Contracts, or in Iraq, "Squeeze PSCs." They are based primarily on a "master development plan" (MDP) that outlines the technical requirements of a given project, including: number of wells to be drilled, production facilities required, equipment and material specifications, and typically a target rate of production to be reached.

The scope of work for the contractor is outlined in the MDP, and once completed, the contractor turns over operations to the National Iranian Oil Company (NIOC). The NIOC takes over responsibility for operations once the contractor departs, and pays back the contractor out of up to 65% (70% in the first contract with Total only) of production. The amount of remuneration is based on capital costs, "bank charges," and a predetermined "fee" that can range from 35% to 50% of the agreed capital investment.

The MDP represents a significant engineering design effort in and of itself. In 1995 Conoco, working with the NIOC, was developing an MDP when the US State Department stepped in and refused to allow Conoco to continue with the contract. Within months, Total stepped in and signed the contract Conoco was hoping to fulfill.

The buy-back contracts virtually guarantee the contractor a minimum ROR, as long as cost projections are correct. These RORs reportedly range from 15% to over 18% in the early buy-backs.

Gas comes of age

- **Namibia** – 30+ year old gas fields get a pipeline
- **Mozambique** – stranded Pande and Temane fields are piped to South Africa's Sasol power plant
- **Sakhalin II** – gas pipeline becomes the first in history to be constructed without long-term contracts

- **Alaska** – a US$20 billion gas pipeline is to be largest project in the Americas
- **Wyoming** – Coal Bed Methane project 15+ years in development
- **Texas** – Barnet Shale development
- **Black Sea** – the US$3.4 billion Blue Stream is most sophisticated gas pipeline constructed to date
- **OGEC** – an Organization of Gas Exporting Countries is being organized.

Mexico's multiple service contracts (MSCs)

Pemex and the Strategic Gas Plan

In 2000 Mexico's state-owned monopoly, Petróleos Mexicanos (Pemex), launched the Strategic Gas Plan, hoping to increase production to 8 billion cubic feet per day by 2008. The company announced five objectives:

1. To increase gas production through MSCs.
2. To diversify gas supply sources from abroad and increase liquified natural gas (LNG) imports.
3. To flare associated gas (Pemex flared an estimated 266 million cubic feet per day in 2002).
4. To expand gas transport, distribution, storage facilities, and cross-border interconnection capacity between Mexico and the USA.
5. To increase exploration funding and proven reserves.

The Mexican Constitution clearly prohibits foreign ownership of the nation's hydrocarbons. This limited Pemex to the use of service agreements, which proved cumbersome.

Mexico's MSCs were designed to streamline development

The MSCs are complex and not particularly attractive to the major IOCs (oil companies never take title to hydrocarbons, and are unable to book reserves). However, the first MSC licensing round for the Burgos basin raised nearly $5 billion.

The MSC contracts, as described by Mexico's President Fox, are simply bundles of various service contracts. By bundling the contracts, Mexico hopes to streamline development of their northern region's gas fields, and reduce their dependence on expensive gas imports from the USA.

The controversy is that the MSCs appear to hand over considerable control to the IOCs. And, opponents claim that they are unconstitutional.

Venezuela tells international oil majors to cut booked reserves

Venezuela's energy minister yesterday told major international oil companies operating in the country to cut Venezuelan oil reserves they booked with the US Securities and Exchange Commission because of changes in Venezuela's oil laws.

Companies such as BP, Total and Chevron will be affected, said Oil Minister Rafael Ramirez. "The new joint ventures make very clear that the reserves are Venezuelan property," Ramirez said. "The Repsol YPF situation [reserves cut] is not an individual one, but affects all the companies that work in Venezuela." (*Greenwire*, Friday, February 3, 2006)

Libya's Fourth Generation Exploration and Production Sharing Agreements (EPSAs)

The EPSA IV feeding frenzy

Competition in the latest Libyan license round – January 29, 2005 – was intense with an average seven bids per block for 15 blocks in five basins. This is an amazing number of bids. The driving forces behind this frenzied bidding included:

- High oil prices
- Availability of prospective acreage that had been off-the-market for years
- Scarcity of opportunities available elsewhere.

In a sealed-bid competition, companies bid:

"M factor" Production multiplier or NOC share of gross production
"B factor" Signature bonus – which was considered in the event that there was a tie in the "M factor" bid.

Block sizes – were just over 2 million acres each, about average for non-frontier acreage.

Work program – averaged $20 million per block, which is fairly light by world standards.

Signature bonuses – averaged $8.8 million per block.

Taxes – Paid in lieu; paid for and on behalf of the IOC by the NOC.

Government take – approximately 88% on average. The terms of the highly contested Block 124 are among the toughest in the world.

Nigeria – Sao Tome JDZ license round

An effort to make it easier for local companies to participate in the second licensing round in the Joint Development Zone came under fire. Many of the local companies that won bids for blocks in the JDZ simply did not have the experience or financial backing required to follow through with contract requirements.

African petroleum producing nations look to harmonize terms

The African Petroleum Producers Association (APPA) met in Gabon in January 2005 to discuss harmonization of fiscal terms. The largely unexplored continent has huge potential and is organizing to understand and maximize the huge potential.

Conclusions

In the world today, as before, there is significant competition for a limited amount of exploration capital. At the same time, exciting acreage is hard to come by. If governments want to increase exploration activity in their countries, they have to offer terms commensurate with their geological potential, location, and political situation. Those terms must also meet the needs of the country – poorer countries need to pay closer attention to how they will receive potential revenue. They simply cannot wait years to receive a share of profits. The trade-offs can be daunting. Important as fiscal design is, the means by which governments choose to *allocate* acreage or projects is just as important. When governments have good geology they are more likely able to allow companies to "bid the terms," as they did recently in Libya. However, with less exciting prospectivity they will probably have to design terms themselves and allocate licenses in a user-friendly way. Otherwise they are likely to be disappointed with the level of exploration activity in their country – a common complaint.

It is difficult for governments to keep oil companies and citizens happy simultaneously. This is where transparency can have a dramatic impact. It is part of the education process and one of the best ways to control expectations and promote a healthy business environment.

Notes

1. Catholic Relief Services and Bank Information Center-PFC Energy, *Chad's Oil: Miracle or Mirage, West African Sector: Oil Value Forecast and Distribution*, December 2003.

2. Ibid.
3. "The Global Market of the 1990s" [with Libya 2005 EPSA IV terms], *Oil and Gas Journal*, April 18, 2005.
4. A view expressed by World Bank economists during country studies in Bolivia in 2003–04.
5. D. Johnston, *International Petroleum Fiscal Systems*, PennWell Books, 2001.
6. Quoted on website, <www.akdi.ru/sf/akdi.htm>.

3
Contractual and Fiscal Outlook for E&P in Latin America

Carlos A. Garibaldi

Latin America is a rather modest contributor to global oil and gas supply, with two globally relevant resource countries (Mexico and Venezuela) and several countries that happen to have hydrocarbons. With the exception of very few plays still in the super-major league, the rest seems to have been left behind as interstitial niches for junior majors, international and indigenous independents, and transnational state oil companies (NOCs).

Politically, after the perceived failure of liberalization policies of the 1990s, the region has regressed in large part to an eclectic populist trend. In response to some of the stresses affecting our industry (higher rents created by the demand-driven price windfall, psychology of scarcity, reserves concentration, petro-nationalism, company consolidation, NOC globalization, etc.) several local governments have adapted their E&P contractual and fiscal regimes to their new perception of the market. They seem to have chosen their positioning between bipolar directions, and their new stances may or may not be aligned with the industry's perceptions of their prospectivity and materiality.

This chapter briefly visits all petroleum countries from Mexico to Argentina, while assessing and benchmarking their perceived prospectivity combined with their contractual and fiscal trends. It pays particular attention to recent developments in Venezuela, Colombia, Ecuador, Bolivia and Argentina.

Introduction: Global E&P industry environment

"Strategy failures are generally failures in confronting reality."[1] This is as valid for host governments as it is for exploration and production (E&P) companies. A successful national hydrocarbon strategy necessitates monitoring and anticipating external trends, so that policies and contractual/fiscal terms and incentives can be adjusted and properly realigned.

The current external environment is rapidly evolving and the "force field" is becoming more intense (higher degree of rivalry, higher opportunity "supplier" power for the hosts, etc.). The most notorious and relevant phenomena are: (1) higher rents created by the demand-driven price windfall, (2) a looming medium-term scarcity of resources, setting the stage for a potentially severe supply/demand imbalance that will sustain or boost high hydrocarbon prices, (3) the resulting imbalance between cash and available opportunities, (4) the recent consolidation of super-majors and the logical upgrading of their materiality needs, (5) the raising of *internal* barriers – only the select few may aspire access to significant conventional reserves located in technological, contractual and/or geopolitical frontiers, (6) the eruption of transnational NOCs following strategic mandates (and not necessarily at our standard cost of capital criteria) as aggressive competitors, (7) the rebirth of petro-nationalism motivated by the current price windfall, (8) costs rising and seeking their share of the bigger pie, and (9) an incipient general toughening trend for contractual and fiscal terms.

While energy demand grows, fueled by the compounded vegetative and economic growth, the legendary global supply capacity peak seems now within reach. According to Matthew Simmons,[2] the world's 14 largest fields based on *current* output, account for 20% of total petroleum production and are on average 47 years old.

Dissatisfaction with recent exploration results is rampant and super-majors are failing in their organic growth. Exploration results are relatively poor (having peaked in the 1960s), and contribute just about 50% of global reserve replacement. Finding and development (F&D) capital costs per barrel are now in the order of $8–10 per barrel of oil equivalent.

Since majors are already vertically integrated and conglomerates' attempts at diversification have always failed, a logical strategic response is to consolidate in order to at least enjoy further economies of scale. However, these eventually exhibit diminishing returns too. If they cannot deliver on production rate growth, super-majors should be aiming once again at return on capital employed (ROCE). This drives them to focus on areas of high materiality at low unit costs (e.g., Persian Gulf, holding two-thirds of the world's remaining conventional crude reserves, the Former Soviet Union, West Africa, etc.), and to stop seeking growth or even eventually to withdraw from areas of low materiality and high unit costs. Portfolio rationalization via divestment is not yet occurring (companies still cling to any source of cash flow that they have), but eventually it will.

E&P environment in Latin America

Latin America and the Caribbean represent 11% of the world's oil reserves, 13% of its production and 8% of its consumption. As happens globally, governments control the vast majority of reserves (notably in Mexico, Venezuela, Ecuador, Colombia and Brazil). Over 80% of these reserves are found in Venezuela and Mexico, while the rest is split mainly among Brazil, Argentina, Colombia, Trinidad and Tobago (T&T), Ecuador and Bolivia.

Mexico is the region's top oil producer with 3.5 million barrels of oil per day, followed by Venezuela with 2.6 (3.1 according to its government), Brazil with 1.8 and Argentina with 0.8. All four, along with Colombia and Ecuador, are net exporters.

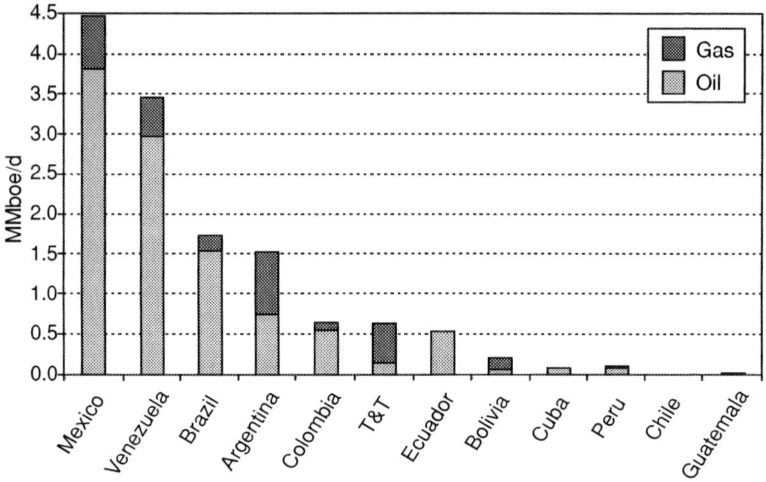

Source: Adapted from BP; Petroleum Economist, LatAm Energy, IEA, Business Monitor

Figure 3.1 Latin American Oil Production in 2004

Both importers and exporters are dependent on oil and vulnerable to price oscillations. Past crises were surmounted by borrowing, but new credit for the region is tight.

As stated above, the confluence of external forces has resulted in industry consolidation. Super-majors may have painted themselves into a corner by falling into the materiality trap. To feed their now even huger appetites, they must seek larger targets that usually lie in the technological and geopolitical boundaries.

In our hemisphere, the future "elephants" are expected to be in Alaska (subject to environmental and political constraints), the US Gulf of Mexico (GoM), Mexico (not yet open for contractual reward for risk), Venezuela (showing modest new exploration and rehabilitation yields in conventional plays, very interesting heavy oil and gas opportunities, and a hostile business environment), and offshore Brazil (a technical disappointment for the IOCs thus far).

Elsewhere, the perception of prospectivity and materiality in Latin America is relatively modest. Technical risks are neither better nor worse than elsewhere. Surface risks (political, environmental, security) tend to be higher. Contractual and fiscal terms are as tough as anywhere else. Significant reserves may be hard to monetize because of geographic challenges, and lack of infrastructure or suitable markets. In steady state, oilfield services cost much more than in the USA.

In conclusion, Latin America is not *very* interesting from the ROCE standpoint (prospectivity x profitability). Super-majors seem to be harvesting or leaving most non-glamorous plays, leaving interstitial niches for others. The hosts' client base is gradually shifting to transnational NOCs, junior majors, international independents, and if the governments take the proper measures, national micro-independents.

Independents have higher costs of capital (both equity and debt), are not vertically diversified along the value chain, and are not as diversified geographically either. Thus, they are more vulnerable to surface risks and require higher IRR (internal rate of return) premiums.

Transnational NOCs run with "unfair" competitive advantages. They often do not follow cost of capital criteria but obey strategic mandates instead. However, in some cases they still lack major project management, environmental and sociological competencies and being self-contained, they do not create a large economic multiplying effect.

The "welcome" mat is often gone

Latin America is just recovering from the 2000–01 crises that launched it into political and economic disarray. Following the US-centric "Washington Consensus," market-oriented policies in the 1990s produced a boom by focusing on orthodox monetary policy, often resorting to explicit or tacit dollarization. Regrettably, these measures were not complemented with sound fiscal policy (the painful part), and were marred by corruption and cronyism. While the middle class collapsed, the chasm between rich and poor widened, and public trust in free market policies was lost.

The reaction came in the form of a wave of populist elected officials, ranging from pragmatic socialism through amorphous nationalism to nostalgic "*soixante-huit*-ism." This movement risks disengaging individual countries from the global community and reinstating ideological, class, and ethnic struggles. Some neo-populist leaders confined themselves to electoral rhetoric and symbolism while maintaining economic discipline afterwards; others brought an about-face in economic policy and drastic (and often retroactive) changes of the rules of the game, alienating foreign investors.

Meanwhile, the USA after the attacks of 9/11 saw the Middle East as their geopolitical center of gravity. Latin America (the "backyard") has been neglected, apart from efforts to combat the drug trade and its linked criminal activities.

Alliances among the new populist regimes in the region may initially seem appealing in their attempts to confront the trade and military dominance of the USA, but they must breach daunting geographical barriers, surmount the lack of proper infrastructures and work with populist governments, which, due to social pressure, may be focusing on ameliorating domestic ills, thus seeking short-term improvement to the parts rather than the long-term benefit of the whole.

To make matters worse, the region is perceived as suffering from endemic corruption. With the honorable exceptions of Chile and Uruguay (not surprisingly non-petroleum countries, as ours is perceived as a highly corrupting industry) Latin America is rated very poorly by Transparency International. Although such poll-based surveys can be highly inaccurate, they do reflect general perception, which is often more powerful than reality. For example, Colombia's poor rating is misleading, since general perception may blame the ills associated with drug trafficking, guerrilla warfare and paramilitary activities on a government that is actually trying to stamp out these problems.

Host choices and constraints

The hosts' dilemma is whether to use the oil and gas (O&G) industry as an engine for economic recovery (long term), or to milk it in order to respond to urgent social issues (short term). But in making their decision, they are constrained by boundary conditions.

For one thing, they must still compete with other global oil provinces with better ROCE. It takes foreign investment to grow their energy industries, which in turn fertilize most other industries. Thus, they must start by attracting energy investors. In doing so they must discern between maximizing rent *take* (a dimensionless number) and rate *capture* (money). They must award their asset rights to those who value them most, and will add the most value, not destroy it. They must develop effective tax systems and close loopholes to achieve a fair rent split. But most fundamentally, they must make the desired economic growth possible and sustainable by having a comprehensive energy plan.

A tour of the neighborhood

We will now briefly highlight the individual status of the main petroleum countries in the region, going from North to South. In general, it is still not an overwhelmingly appealing prospect.

Mexico has a spectacular and still untapped prospectivity. The state-owned Pemex enjoys an absolute E&P monopoly, and any contracting mode that rewards risk is unconstitutional.

The multiple service contracts (MSC) for existing gas fields in the Burgos basin are theoretically pure service contracts, but hide a subtle risk component. Nevertheless, this process was charged with "not going far enough" by potential bidders from the USA (personally, I am amazed they went that far without modifying the Constitution first). The resulting population of bidders was idiosyncratically dominated by Latin companies. The later gas MSC round offered only slight improvements.

There are plans to extend a MSC-type model to marginal oil fields. This may prove more difficult because oil has stronger emotional and political connotations. In fact, a round announced for January 2005 was soon derailed by disagreement within Pemex itself. The would-be "marginal" oil field had 10 billion barrels of oil in place, and had accumulated 1 billion barrels of oil. It had been severely neglected since the 1930s, with only sporadic pulling operations since then and is still producing about 5 million barrels per day with a 50% watercut.

The issue here is that the "food chain" that transfers fields from larger companies that follow the "80/20" rule to smaller ones, and then all the way down to micro-independents, cannot function in a country sporting a state monopoly on E&P. By way of illustration, Figure 3.2 shows well density in a single basin divided down the middle both by a river and by two diametrically opposed contractual systems.

One side was developed by the market with a complete food chain, while the other by a state monopoly with 100% government take on rent. It comes as no surprise to learn that government captured more rent.

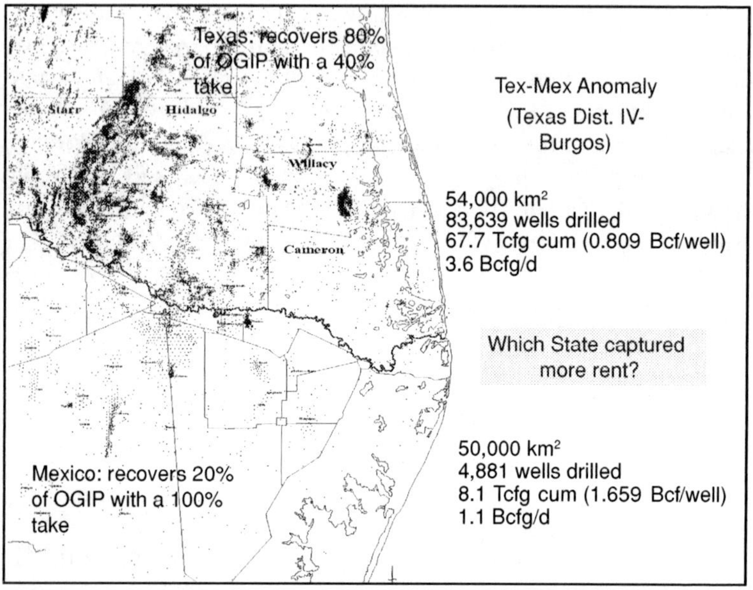

Source: Scotia Group

Figure 3.2 One Basin Bisected by a National Boundary and Extreme Opposite Contractual Frameworks

Mexico still carries "baggage" for their harsh experiences with the old majors prior to President Cárdenas' nationalization. After suffering a defeat in local elections, the ruling party (PAN) won the presidential elections by a razor-thin margin. Runner-up López Obrador (PRD) challenged the results in court, but Calderón was confirmed. The PAN, although ideologically market-inclined, has failed thus far under Fox to push the necessary reforms. Calderón winning 36 percent of the vote in a three-way race is not a mandate either.

I believe in the law of gravity. Mexico needs substantial E&P investment. Any intent to get too "creative" and look for technicalities and loopholes to get around the spirit of the Constitution is likely to be short-lived. Someone in the near future will have to take the bull by the horns, reform the Constitution and allow foreign investment into the country.

Guatemala's prospectivity is low, and it produces mainly low quality asphaltic crude. Unjustifiably tough contractual and fiscal terms, coupled with logistic and security concerns, mean that the country elicits only occasional interest from small "niche" independents.

Cuba enjoys modest prospectivity, although the deep waters of the Gulf of Mexico may yet prove interesting. Since the collapse of the Soviet Union Cuba has had a strong incentive to develop its domestic reserves, because its export commodities are low priced. However, it now enjoys a new patron (Chávez's Venezuela) that has saved the day. Most investors are Spanish and Canadian O&G companies, since US-affiliated firms are forbidden to engage in any activity by the Helms-Burton Act. Curiously, Cuba is the country on which the USA has imposed the most sanctions (more than Saddam's Iraq, North Korea, or Libya).

Trinidad and Tobago have very modest onshore oil potential, but excellent offshore gas potential (about 40 trillion cubic feet (Tcfg) yet to be discovered). While their southern neighbors were arguing, they have been doing; their LNG project is online (five trains) and growing. Its contractual and fiscal terms are rather cumbersome, yet reasonably attractive. The food chain is functioning, with Repsol-YPF recently acquiring BP's mature offshore oil fields (Teak, Samaan, and Poui).

Venezuela has tremendous prospectivity, yet it saved the best for Petróleos de Venezuela (PdVSA). The *apertura* brought majors and independents alike into a feeding frenzy, despite the harsh contractual and fiscal terms. The bid rounds in the late 1990s were a textbook example of herd instinct, hubris, and testosterone-led bidding. Well, our industry has never been accused of being overly rational.

In the meantime, PdVSA had become strategically independent and was engaged in a remarkable foreign downstream expansion. President Chávez shut down further *apertura* and tried to rein-in PdVSA to make it subordinate to his vision. The ensuing governance crisis escalated into a full-scale insurrection on the part of the company, including sabotage.

Surviving this coup, Chávez ravaged PdVSA's management team and staff (expelling 18,000 out of 32,000). PdVSA is slowly recovering its production capacity and recuperating from the talent drain, but is now under heavy political control. Moreover, funds normally earmarked for reinvestment in PdVSA (at least $2 billion) have been diverted without democratic scrutiny to social, political and regional subsidies (servicing Cuban, Ecuadorian, and Argentinean debt), and "Bolivarian Integration" initiatives (Petrocaribe, Petroandina, Petrosud and now, Petroamérica). Royalties have been increased to 30% for oil and 20% for gas, as well as income tax. There is an ongoing forced conversion of Operating Agreements into Joint Venture Companies (PdVSA's working interest is certainly greater than 51%, perhaps as high as 80%).

Having said all this, Venezuela still offers significant not yet monetized gas opportunities in Plataforma Deltana, Falcón, and the Golfo de

Venezuela. There are significant opportunities for tar sands too. And as we know, materiality is our industry's primal motivator.

The Chávez government did not seem at first totally unfriendly towards large IOCs and direct negotiations were common (no tenders and less competition). Companies did not seem overly unhappy with the referendum results that validated Chávez's government with almost 60% of the popular vote. One even announced a major investment only a few days before the referendum. But now Chávez is cracking down on their alleged tax lapses and strengthening PdVSA's grip on revenues.

Colombia offers reasonable prospectivity for independents, because material results in the super-major league (i.e. foldbelt exploration) have been elusive of late. However, thanks to its contrarian strategy, it may well turn out to be the most attractive country in the region for E&P investment.

The Colombian government has taken a brave tack to attract investors, with the creation of the national hydrocarbons energy department (ANH) to administer contracting rights independently from Ecopetrol, and the elimination of the Association Contract that butchered materiality and exploration ROCE. In essence, it has returned to a concessionary system with very moderate royalties (5–25% as a function of type of hydrocarbon and rate, ringfenced by fields within a contract block). The government still captures price-induced windfall rents (30% of excess revenue over an adjustable trigger).

Moreover, the government may eliminate the 7% profit repatriation tax soon, thus bringing the effective income tax rate down to 35%. But although Colombia is signatory to the proper treatises, international arbitration is not acknowledged in the contract. However, a fiscal stability law has just being passed by Congress.

As further attractions, the government maintains a mixed system of formal tenders and direct negotiations (first come, first served). Furthermore, it has created a fund to assist domestic micro-independents both technically and financially.

There is hardly any political risk in Colombia. The country has a stellar tradition of political stability, judicial independence, and continuous healthy economic growth and enjoys a mature democratic process that results in an alternation in power of the conservative and liberal parties.

The pending potential big problems in Colombia are personal and facility security, still threatened by the powerful drugs barons. However, the trend now looks positive. Both leftist and rightist guerrillas have lost the military initiative to the legal forces. Thus, Colombia seems to have reached a turning point; the incidence of pipeline bombings and kidnapping incidents is dropping sharply and consistently.

Security issues still have an adverse influence on exploration and development costs, which may be more than triple those of similar operations in other parts of the region. This affects the net prospectivity (net reserves given a capital effort) and thus, the ROCE.

The ANH is focused on "heating up" frontier basins and taking a marketing approach to promotion. In my opinion they take the "heating" a little too far, since the agency will be spending about $80 million per year in frontier exploration (onshore and offshore regional seismic and stratigraphic wells). This effort requires adequate management and risks the agency, still with severe human resource shortages, growing into a large bureaucracy. Maybe strong fiscal incentives for the frontier basin pioneers could have done the trick.

Perhaps as consolation for losing its licensing role, Ecopetrol has been authorized to expand internationally too. To do so, it would require the creation of a more agile international vehicle.

Ecuador offers substantial prospectivity, and used to be pipeline capacity-constrained. Once the Oleoducto de Crudos Pesados (OCP) was completed, it developed the reverse problem – excess capacity. Together with the ship-or-pay obligations of OCP participants, this should have led to a boost in exploration activity. Furthermore, the south is essentially unexplored for two main reasons: lack of infrastructure and a past border conflict with Peru, recently solved.

The exploration boom has not happened, partly because companies have to wait until the government gets its act together and finally calls a tender that is not aborted for political reasons, and partly because contractors do not make partial area relinquishments as they should. There have been plenty of acquisitions in the recent past, but the "food chain" does not work for exploration. In fact, there are no country-wide tax synergies for exploration; ringfencing is by contractual block for income tax purposes.

Petroecuador's core fields are severely underdeveloped and underexploited and require an urgent and significant infusion of new capital to arrest their decline. There should be significant inter-field potential too. It seems almost criminally negligent that, in an era of high oil prices, this should not have been developed. Meantime, these noble giants – Shushufindi, Libertador and Sacha – reach bubble point reservoir pressures and water breakthrough, making the damage irreversible and sharply decreasing the productivity of any intervention.

Despite Ecuador's exploration and rehabilitation potential, political uncertainty and an ambiguous attitude towards foreign capital dampen its attractiveness. Every rational initiative is aborted by government intervention. Energy ministers and Petroecuador CEOs rarely last more than six months, but seem to resuscitate every few years and walk through those revolving doors again.

The government's stance towards O&G investors appears hostile. For example, there is a perception that Oxy may be "punished" (by having their contract on Block 15 terminated on bureaucratic technicality) because they have just won a VAT arbitration award against the state.

There is also no apparent coherent sector policy. In 2005, the caretaker president, Alfredo Palacios, announced a "Hydrocarbon Plan," that included the wholesale thawing of frozen bid rounds (marginal fields, exploration, heavy oil, etc.). His announcement was received with skepticism by the oil industry.

Royalties are 12.5–60% as a function of rate and API gravity,[3] but some recent contracts have adopted a new function of price and cumulative production, which is obviously more in synch with the current price environment. Some of these new contracts use real prices, some nominal ones.

Ecuador has always had a poor tax enforcement system and needs to close its tax loopholes and provide clarity and stability if it is to attract and retain foreign investment. There are ways of minimizing the 15% labor tax on earnings by creating a staffing intermediary, lowering the income tax by loaning oneself capital from an offshore vehicle and deducting the self-paid interest, etc. If it cannot achieve a functional tax system, Ecuador may have to resort to safer channels of government take, such as the price-progressive royalties.

Not surprisingly, there are pressures to raise the minimum royalty in the Hydrocarbon Law. Moreover, the government has just announced its invitation for all contractors to "voluntarily" renegotiate their contracts to provide a higher rent split in the state's favor. Most likely, they will make the sliding scale a function of price.

Petroecuador is allegedly looking at PdVSA for technical and financial assistance. Moreover, because of spill-over from Plan Colombia, security is deteriorating. While the previous government was aligned with USA's interests in Colombia, President Palacios (who had been vice president until the coup d'état in 2005), has taken the country some distance away.

Peru's oil and gas industry is almost completely privatized and has the distinction of offering the best contractual and fiscal terms in the region.

Sadly, its prospectivity, particularly in the northern Marañón basin, has been disappointing. In spite of this, seven new exploration contracts have been signed since the 2003 stimulus. Industry activity is dominated by Pluspetrol and its Camisea project. Future growth seems geared towards increasing gas reserves in its vicinity.

The present government is weak, so the political outlook is somewhat uncertain, despite the solid economic footing President Toledo inherited from the Fujimori era, and which to his own credit he has preserved. Security could be deteriorating and Shining Path seems to be regaining momentum after some re-engineering.

There is an unsettling rumor that the government may be planning to re-launch Petroperú as an integrated company, likely following Argentina's and Bolivia's example.

Brazil is perceived as offering world-class prospectivity in its deep continental margins. Petrobras and the ANP succeeded in attracting super-majors and significant independents to its initial rounds yet seem to have saved the best tracts for Petrobras. There is disillusionment after the initial (and customary) feeding frenzy.

The fiscal terms are relatively attractive yet very cumbersome (there are about 62 different taxes to manage). Brazil has not perniciously revised its contract terms and remains contractually stable. Petrobras, with remarkable deepwater technological capacity, is expanding internationally. President Lula's government's ideological roots could have mandated a different role for a state company, yet nothing has changed since his election in 2002.

Bolivia has little oil prospectivity, but has managed to stimulate gas reserves well above the available market's capture capacity (about 35–50 trillion cubic feet). The focus then shifted from gas exploration to gas monetization. The gas exports are dominated by Brazil, yet flow rates through the Bolivia-to-Brazil pipeline are still below expectations, as Brazil has failed to deregulate its electricity. Brazil now and then exerts its buyer power when renegotiating the gas price, while gas is being discovered in Brazil itself. However, an energy crisis has forced in Argentina, Bolivia's neighbor to the south, to acquire (and become dependent on) Bolivian gas once again.

The Pacific LNG export project was killed by historical, political and cultural issues, as a past government failed to decide on an export harbor (Peru or Chile). Using popular hostility towards Chile as a cover, leftist indigenous forces succeeded in derailing the project and in bringing down Sánchez de Lozada's and Mesa's governments. These administrations were particularly bland and any concession offered to appease the opposition was greeted as a sign of weakness. Leftist leader Evo Morales, of the coca growers' union, has been elected and is in the process of changing hydrocarbon regulations.

Domestic supply obligations at regulated prices and the forced renegotiation and conversion of existing contracts with hardening of the terms, are additional causes for investor despair. The new Hydrocarbon Law introduced a Hydrocarbon Tax (IDH). This is actually a 32% royalty that springs from the same basis as the 18% royalty and provincial participation payment: sales. Additional income taxes total 34.4%

Existing contracts must be forcibly converted to one of three contractual models that take over after the 50% royalty burden: (a) an Operations Contract: providing yet another override (albeit likely to be very modest) to YPFB, (b) a Production Sharing Contract: the same, but with slight accelerated cost recovery, and (c) an Association Contract: like Operations but with an option for YPFB to back-in with working interest after reimbursing successful wildcats with only nominal US dollars. This is worse than the model just abandoned by Colombia, and obviously inapplicable for existing fields.

Actually, projects are effectively eliminated by the application of a 50% royalty burden. This implies that there cannot be any new exploration, even for 4 Tcfg targets. Small fields should be devastated, even with up to 16% IDH rebates (in the Law, but in clear contradiction with a prior article in the same Law).

The economics for continuing development of the giant existing fields (needed to meet Brazil's and Argentina's needs) are barely marginal. Three out of the four giant gas field operators (Repsol-YPF, Total, and BG) have already triggered their international arbitration mechanisms.

Chile, although a macroeconomic success, is still a net hydrocarbon importer, with modest and declining domestic production. As it cannot trust Argentina as a reliable supplier and is refused as a natural client by Bolivia, it must build its own LNG receiving terminal.

Curiously, it maintains a state monopoly (ENAP) that operates internationally as Sipetrol, and is quite active in the region as well as in North Africa. Governance issues derived from creating a separate company under private law, that diverged culturally, resulted in Sipetrol's virtual reinsertion into ENAP.

Argentina is a mature petroleum province, with moderate prospectivity, good infrastructure and mature domestic oil and gas markets. Technological and conceptual paradigm shifts are yielding positive exploration and exploitation results in the north flank of the Gulf of San Jorge basin, but this seems to be the only bright spot. There is some high cost-risk-materiality exploration potential in the deeper continental margin, but its appeal to super-majors (its natural candidates) is marred by long-term uncertainty and the currently hostile investment climate.

Like Bolivia, Argentina had been a poster-child for privatization and deregulation in the E&P industry during the 1990s, offering hugely attractive terms and in consequence, succeeding in augmenting production by 70% while increasing oil and gas reserves in the same proportion. As a secondary effect, a cadre of nimble local independents blossomed and expanded internationally, only to be later mostly consolidated or acquired.

After a severe political and economic crisis in late 2001, Duhalde's government changed the rules, damaging the industry's perception of reliability, credibility and risk. An export royalty ("retention") was imposed but was countered by a severe devaluation that managed to maintain profitability for oil. The government in fact clinched the rent windfall created by the cost devaluation. The net-backs remained basically constant, and profitability even improved slightly because of the timing issues (costs are cut earlier and retentions are paid later). This acted as an incentive to invest in quick-response development but to defer longer-term exploration (which was wise, because costs are making a comeback). The same devaluation crippled the gas industry, since domestic gas prices and electric tariffs were *pessified* and controlled (really a rent transfer from producers to consumers). Reserves for both oil and gas are dropping and hydrocarbon self-sufficiency for the near future is challenged.

President Kirchner's response to this last instance has been a hardening of fiscal terms (increased export retentions, really an additional 14–34% royalty), bringing the government's take on rent to about 70%. Agreements and commitments to Chile have been violated and the government is recreating a new NOC (ENARSA), whose main mission seems to be to pioneer frontier exploration in deep waters. Recent tax incentives only apply in case of association with ENARSA.

Having said all this, the government seems ripe for a post-electoral reality-check and maybe a change of direction. The country has basically recovered its industrial capacity at an incredible growth rate (9%) but according to some critics, this is just a "dead cat's bounce." Further growth will require fresh direct domestic and foreign investment. This implies changing the hostile discourse against foreign companies (which may have been no more than an electoral stance) and providing sufficient energy for a growing industrial base. Again, I believe in the law of gravity.

Conclusions: A bipolar region?

Our region is not particularly attractive when we take an overview of perceptions of prospectivity and materiality, contractual and fiscal frameworks, political and fiscal stability, and personal security issues. Is there an "all green" benchmark in the area at all? In my view yes there is, the US Gulf of Mexico.

Source: Harrison Lovegrove

Figure 3.3 Regional Attractiveness

In response to some of the global stresses affecting our industry, several local governments have adapted their E&P contractual and fiscal regimes to their new perception of the market. They seem to have chosen one of two extreme positions, as can be observed in Figure 3.4, which benchmarks typical crude (only) targets. Some repudiate their policies of the 1990s, and even appear hostile to foreign (or private) investment (Venezuela, Argentina, Bolivia, Ecuador), while others do the opposite and offer strong incentives to lure investors (Colombia, Peru).

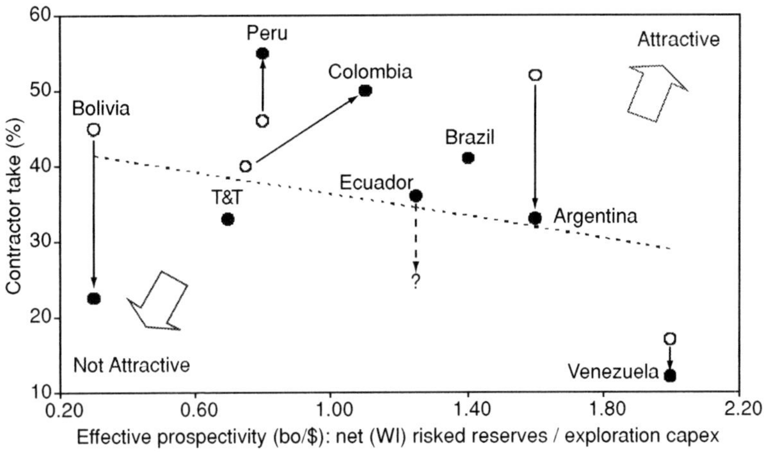

Source: Scotia Group/Harrison Lovegrove

Figure 3.4 Regional Competitiveness (for crude only)

While Mexico and Venezuela should both be able to exact the lion's share of petroleum rent should they decide to open up or reopen to foreign investment, the host governments of the countries with less attractive prospectivity should maybe rethink their current positioning.

Their new stances may or may not be aligned with the industry's perceptions about their relative attractiveness. They risk missing out this cycle and incurring tremendous opportunity costs. Sadly, these errors will not be measured and punished.

Notes

1. Quoted from Joan Magretta, in J. Magretta and N. Stone, *What Management Is: How It Works and Why It's Everyone's Business*, New York: Free Press, 2002.
2. M.R. Simmons, *The World's Giant Oilfields*, Hubert Center Newsletter #2002/1, M. King Hubbert Center for Petroleum Supply Studies, Colorado School of Mines, 2002.
3. API gravity – a system of classifying crude oil based on its specific gravity, whereby the greater the gravity, the lighter the oil. Source: *National Association of Royalty Owners (NARO) Oil and Gas Glossary*, April 7, 2006, <http://naro-us.org/resources/glossary/a.htm>.

4
Upstream Policy Responses – A Norwegian Perspective

Gunnar Gjerde

The last couple of years have seen oil prices – spot as well as futures prices – at historically high levels. Oil demand has grown faster than production capacity, causing OPEC's reserve capacity to shrink to historically low levels. Recent development in the oil market signals a need for increased upstream investment. Industry and governments have a shared responsibility to expand production capacity. This chapter will address Norwegian policy responses to increased activity level on the continental shelf, as well as developments in the Norwegian petroleum sector in the last few years.

Norway is still a promising petroleum province with large resources to be discovered. Slightly less than one-third of the recoverable resources have been produced.

We have recently increased the frequency and scope of licensing rounds, as well as tightening work commitments. The latest forecasts indicate a high level of exploration activity in coming years. For 2005, twice as many exploration wells were expected to be drilled as in the previous year.

An important priority has been to establish the Norwegian Continental Shelf (NCS) as an attractive province for new entrants. In the past six years, more than forty new companies have become pre-qualified or have established operations on the NCS. The new companies will not substitute, but will supplement the existing players.

Efforts to increase the recovery rate from existing fields are high on the agenda and an important element in the Norwegian government's strategy to sustain the oil production level. Increased recovery rate requires targeted efforts by the industry, also with respect to research and development (R&D). Today we expect an average recovery rate of 46% on the NCS. The Norwegian government's ambition is to reach a recovery rate of 50% or even higher.

Looking at field developments, activity is picking up. The investment level in 2005 was expected to reach an all time high.

Norway is now producing close to 3 million barrels of oil per day and 85 billion cubic feet of gas annually. For oil, the challenge is to maintain current production levels as long as possible. Gas resources are plentiful and production is expected to increase steadily in the years to come.

Norway is the main exporter of gas to Europe and natural gas is gaining in importance in our petroleum activities. With a favorable geographical position close to the gas markets in Europe, we also have the largest offshore gas transportation system in the world, which facilitates delivery to these markets.

Currently, we supply almost 15% of European consumption. Security of gas supply is an increasingly important issue in Europe, making it crucial that Norway remains a stable, secure supplier of gas.

Development in the Arctic region could change the overall prospects on the NCS significantly. Many believe that the world's Arctic regions contain significant amounts of petroleum resources. Despite the environmental challenges, the government is strongly in favor of increasing exploration and development in the Barents Sea and the region will be included in our next licensing round.

Introduction

In the late 1950s, there was no oil industry in Norway, and very few people believed that the Norwegian Continental Shelf concealed rich oil and gas deposits. In the 1960s, however, international oil companies discovered the potential of the NCS as a petroleum province. With the Ekofisk discovery in 1969, the Norwegian oil adventure began. Production started in 1971, and in the following years a number of major discoveries were made. During the development of these fields in the 1970s and 1980s Norwegian oil companies were established and a supply industry was developed.

Today, there are around 50 fields in production on the NCS. Production from these fields corresponds to around 20 times the domestic consumption of petroleum, and has established Norway as a key supplier to the global oil market and the European gas market.

Unlike many other petroleum provinces, self-sufficiency has never been the most important issue for Norway, because the vast majority of our petroleum is exported. Our goal has always been to maximize value creation to benefit Norwegian society. National management and control, building a Norwegian oil community and creating an attractive investment climate have been key goals for Norwegian oil and gas policies.

An attractive investment climate must provide a stable and predictable framework for investors. At the same time it is in the interests of both the authorities and the investors that the framework is flexible enough to respond to changes in society, the markets, industry trends and the development of activity in the oil and gas fields.

It is fair to say that in many ways we have succeeded both in providing stability and in adjusting our policies over time. So far around $300 billion have been invested in exploration, field development and transport infrastructure on the NCS. We have been able to attract investments from the largest international companies and to build a strong Norwegian oil industry that can compete on equal terms. We have also responded to society's demand for increased focus on safety and environmental issues.

Upstream policy responses

The two most important factors for future activity on the NCS are the oil price and the size of the resources left to develop.

In the last couple of years we have seen oil prices – spot as well as futures prices – at an all-time high. Worldwide oil demand has grown faster than production capacity, reducing OPEC's spare capacity to historically low levels. There are reasons to believe that in the long term oil prices will remain fairly high. Although we have seen some changes in consumer behavior, a robust outlook for demand is anchored in the transportation sector and economic growth in Asia. There is evidence to suggest that exploration and developing costs outside OPEC will continue to increase, and mature non-OPEC regions are not able to increase production. All this leads to high prices for long-term crude futures contracts and high long-term price estimates from most consultants.

The growth in demand for oil and gas has strengthened the need for increased investment on the NCS. High investments are a crucial condition for exploiting as much as possible of the resource base. The investment level in Norway is at an all time high: in 2005 it reached about $14 billion and is expected to stay high for roughly the next four years; thereafter it will gradually decline as the largest field developments are completed.

The total recoverable oil and gas resources on the NCS are expected to be 13 billion standard cubic meters (scm) oil equivalents (oe), of which one-third has already been produced and sold. The figures are uncertain, and vary between a high estimate of 16.3 billion scm oe and a low one of 10.6 billion scm oe. Continued exploration and more discoveries are preconditions for producing the remaining resources, as are efforts to increase the recovery rate from existing fields.

It is a challenge to make sure that activity on the NCS remains on a high level in the coming years, and that our remaining resources are developed. As parts of the NCS are maturing and we are to a larger extent moving in to deep water and Arctic areas, it is important that we remain competitive. We have based our system on participation from international companies and must constantly focus on having a system that will attract capital and the best companies.

In Norway framework conditions have always been adjusted to new challenges in a close dialog with the companies involved. We aim to be world leaders in competence, flexibility and competitive framework conditions. In this chapter I will focus on policies related to exploration, new entrants, the Arctic region, infrastructure and increased recovery.

Exploration policy

Our estimates show that about a third of the remaining resources on the NCS are yet to be discovered. Large parts of the NCS, like the deepwater areas in the Norwegian Sea and much of the Barents Sea, are little explored. Even the most mature parts of the NCS are relatively little explored compared to, for example, the United Kingdom Continental Shelf and the Gulf of Mexico. To uncover the huge values that are hidden on the NCS, extensive exploration activity will be necessary.

In the late 1990s – in part as a consequence of a sharp fall in oil prices and in line with many other petroleum provinces that had been open to the international industry – we experienced a drop in drilling activity on the NCS. Another contributing factor was a series of mergers among the world's largest oil companies and the resulting focus on synergies and cost reductions. The big international companies are looking for giant fields, and this is the main reason why they now show less interest in the more mature parts of the NCS.

Against this background the Norwegian government has responded with several changes to its licensing policy. The industry's access to exploration acreage has been increased and predictability has been improved. The system of licensing rounds has been changed to make a clear distinction between mature and frontier areas.

In mature areas a system of awards in predefined areas (APA rounds) was established in 2003. With this change the industry can apply every year for licenses within areas on the NCS that are close to existing or planned infrastructure. Compared to the previous licensing system, the industry now enjoys increased predictability with regard to the acreage that will be available in the years to come. It also has access to much larger areas than before.

At the same time the conditions applying to new licenses have been tightened, requiring faster exploration or relinquishment of licenses with low activity.

In frontier areas we have had a policy of stepwise exploration. When moving into a new region, the Ministry of Petroleum and Energy picks out key blocks that can give information about a larger area. Based on the results of core wells from these key blocks, other areas are made available at a later stage. The effect of this policy has been that the well-density on the NCS is lower than, for example, on the United Kingdom Continental Shelf, but our rate of success is rather higher.

The two last licensing rounds have been large. As a result of the increased access to acreage and the high oil prices we entered 2005 with expectations of a doubling of the exploration activity compared to 2004, when 17 exploration wells were drilled. We have, however, like many other countries and companies, experienced a new obstacle – namely rig-availability. But although this is a huge challenge, unlike access to acreage the answer lies with the industry.

New companies

As mentioned earlier, the composition of companies on the National Continental Shelf has been a contributing factor to the low exploration activity. In addition to the mergers discussed above, the maturing of the Shelf has created a demand for a new composition of players. At the same time there have been some major changes in the petroleum industry with implications specifically to the importance of smaller companies. Technology that was previously available only to a few companies is now readily available to a broad spectrum. This has led to a significant number of medium-sized companies having access to state-of-the-art technology; these companies have also grown larger, with sound financial backing.

Against this background the Norwegian Ministry of Petroleum and Energy changed its policy towards new entrants back in 1999/2000. The changes allowed smaller companies to establish themselves on the NCS and facilitated this through a system of pre-qualification. It is, however, still a prerequisite that new companies should be able to contribute to value creation on the Shelf by challenging the established player and contributing new ideas. The government does not see access to capital as one of the challenges on the Shelf and therefore we do not see a need for purely financial investors or oil companies that mainly speculate in oil prices.

It is important to underline that the established companies are still essential to the future developments of the Shelf. The new companies are not meant as a substitute to the existing ones, but as a supplement. It is very encouraging to see the interest from established and larger companies in the recent licensing round on the NCS.

In 2004 the Norwegian authorities went one step further and more actively went out to promote the NCS to companies that through their track record in other parts of the world had proven their ability to create value.

The changes made have been welcomed by the international oil and gas industry. Since 2000, more than forty new companies have received pre-qualification or in other ways established themselves on the NCS. With very few exceptions, these companies have become active players and they have especially contributed in focusing on mature parts of the Shelf. Today we also have several companies under assessment for becoming either operator or non-operator licensees on the NCS.

In 2004 Norway also adapted the petroleum tax system to make the NCS more attractive to new companies. The policy changes reduce the risks for companies that enter the NCS with an exploration strategy. With the change, companies that are not in a tax-paying position will be reimbursed the tax-value of their exploration cost annually. This means that new entrants only pay 22 cents in the dollar for their exploration expenses.

In 2004 the government largely rejected a massive claim from the oil companies to reduce taxes. The prices we see now have a greater impact on the companies' income than any tax change would have had.

The Arctic region

According to the US Geological Survey, as much as 25% of the undiscovered oil and gas resources in the world are in the Arctic regions. The Norwegian authorities do not necessarily share this view, but it indicates what potential this large region might have. In light of the anticipated future demand it is clear that the Arctic region can play an important part in the future supply of petroleum.

Entering into the Arctic region does however bring its own very serious challenges. On the technical and commercial side, there are the long distances to the markets, the fact that there is little or no infra-structure in place, and the harsh weather conditions with ice and long periods of darkness. The Arctic is an expensive place to explore and produce petroleum and the discoveries need to be big in order to make investments worthwhile.

There are also environmental challenges to be considered. The Arctic is a very sensitive environmental area that is largely "untouched by humans." Large expanses are covered with ice and contain a unique ecological plurality. Even those areas that are not covered by ice are environmentally sensitive. So far there has been no petroleum activity in the ice-covered areas and any possible future activity will have to be handled very delicately.

The Barents Sea

The Norwegian part of the Barents Sea is by far the least explored part of the NCS. Since the beginning of petroleum activity here in the 1980s, only a little over 60 exploration wells have been drilled. In the rest of the NCS, there are in excess of 1,000. The geology of this area is therefore very little known but its potential is believed to be high.

The Norwegian Petroleum Directorate estimates that about a third of our undiscovered resources are to be found in the Barents Sea, not including the disputed areas on the border with Russia. Estimates of these resources suggest that they are in excess of 6 billion barrels of oil equivalent, though lack of geological knowledge obviously makes this a very rough figure. If new discoveries that prove new geological models are made, the estimated resource potential will be adjusted upwards. The Barents Sea is now the region on the NCS where there is the highest chance of making large new discoveries.

The Norwegian petroleum activities are moving north. The Snøhvit LNG project, which will come on-stream in 2006, is the first development in the Norwegian part of the Barents Sea; there have been some other discoveries, but as yet there are no development plans for them. The most promising of these discoveries is Goliat.

Over the last years we have also put considerable effort into evaluating the consequences for the environment and fisheries of possible petroleum activities in the northern areas.

Given the resource potential on the Norwegian and the Russian sides of the Barents Sea, including the super-giant Shtokman field, this region might in the future become one of the most important petroleum provinces in the world. There is potential for huge synergies if the two sides are developed with some kind of cooperation. Maintaining a good dialog with Russia on energy issues is therefore a high priority for Norway.

Infrastructure

Norway has a very well developed infrastructure. We have the largest offshore gas transportation system in the world, which paves the way for Norwegian gas to be delivered to the important European market. The favorable geographical position close to this market is also of great significance. Norway is already one of the major exporters of gas to Europe – and natural gas is gaining in importance in our petroleum activities: currently we supply almost 15% of European consumption. Security of gas supply is an increasingly important issue in Europe, so it is crucial that Norway remains a stable and secure supplier.

To ensure efficient use of the transportation infrastructure, a new management system was introduced in 2002. An independent operating company, Gassco, was established, operating most of the upstream pipeline network. At the same time, unified ownership of the pipelines was established and a regulated access regime introduced.

As the Norwegian Continental Shelf becomes more mature, it is important that existing infrastructure and licensed acreage is exploited in the best possible manner. To ensure development of profitable discoveries, third parties must get access to infrastructure on non-discriminatory and reasonable terms. The Norwegian Ministry of Petroleum and Energy has therefore established regulations to achieve efficient use of existing facilities in order to ensure good incentives for licensees to undertake exploration and production.

Increased recovery

The petroleum industry and society at large share a common interest in pursuing every drop of economic oil and gas. A quarter of the Norwegian resource base is still to be discovered. In the long term, frontier acreage, such as deepwater areas in the Norwegian Sea and the Barents Sea, is a key factor determining the outlook for the NCS. In the short to medium term, the main task is unlocking the full potential from developed areas. If we are to succeed, a number of issues must be addressed.

Recovery from fields on stream needs to be fully exploited through efficient operations

Incremental reserves are achieved through both cost reduction and enhanced recovery. As we go forward, two thoughts need to be at the top of our minds: reducing costs and increasing recovery. In a high oil price environment there is a strong incentive to increase production. To realize the full potential it is imperative that we remain focused on cost control and developing new ways to operate efficiently.

Can the government contribute? When private companies exploit finite natural resources on behalf of society, it is crucial to have the right players. The government must ensure that licenses are in the hands of companies that are eager, competent and otherwise resourceful to pursue and exploit existing opportunities. To fulfill this, we need a business environment characterized by competition and diversity. To empower diversity we encourage companies to enter the Norwegian Shelf. New entrants alongside established companies provide an environment in which asset owners may develop aligned partnerships to pursue ultimate recovery from the Norwegian Shelf.

At the same time, the Petroleum Directorate will closely follow the companies' plans for increased recovery to ensure that we get as much as possible out of the reservoirs in an economic way.

Discoveries need to be developed

At present, there are 60 undeveloped discoveries on the NCS. For some there is a limited window of opportunity, because utilization of existing infrastructure is the only, or the most, profitable development, option. We need to acknowledge that the extensive infrastructure which has been established so far is a major asset. Hence, decommissioning is not acceptable unless all profitable projects that hinge on such infrastructure have been exhausted.

In 2005, Norway and the UK agreed a Framework Treaty on cross-border petroleum cooperation. The Treaty implies a strengthening of the cooperation between the two countries, the oil companies and the supply industry and it has already facilitated the development of several cross-border fields.

Field life must be extended and exploration near infrastructure ramped up

Prolonging the life of fields on stream postpones decommissioning costs. But first and foremost it generates a new set of opportunities: opportunities to exploit the nearby exploration potential, opportunities to promote and tap into new technologies, and opportunities to maintain Norway's position as a major exporter of much needed energy.

Conclusion

To summarize, I would like to bring your attention to The BIG Five. The BIG Five is the Norwegian Petroleum Directorate's new ambition for reserve growth within the next ten years. The goal is to add 5 billion barrels of oil, exceeding current plans by about 1.5 billion barrels – or the equivalent of one year of Norway's total production at today's record level. This ambition comprises the most important issues looking ahead: adding reserves through exploration, getting discoveries on stream, and adding value from existing assets through efficient operations.

Acknowledgments

The author wishes to thank Anne-Chatrine Aarhus, Executive Officer at the Norwegian Ministry of Petroleum and Energy, for her invaluable help in the preparation of this chapter.

5
Policy Challenges – A Russian Perspective

Jennifer Josefson, Partner, White & Case

This chapter discusses the Russian government's 2005 draft subsoil law which, if adopted, will significantly change the current regime for subsoil exploration and production activities within the Russian Federation. Following its submission by the government in June 2005, at the time of writing this chapter in October 2005, the Draft Law was being considered by the Russian State Duma (the Russian Parliament).

The chapter introduces the political context in which the Draft Law was proposed and considers how it will impact foreign companies and investors as well as discussing the changes it will have on the subsoil regime in Russia.

Introduction

The subsoil licensing regime for exploration and production activities within the Russian Federation is primarily governed by the Subsoil Law,[1] which was originally adopted in 1992 and has been amended many times since. In June 2005, a new draft (the "Draft Law"), which is intended to replace the current Subsoil Law in its entirety, was submitted by the government to the State Duma (Russian Parliament).[2] The Draft Law is a combined work of the Ministry of Natural Resources and the Ministry of Economic Development and Trade and represents a thorough revision of the current legislation; if adopted, it will result in a substantial overhaul of the current regime governing subsoil use.

The Draft Law has been introduced in a climate of political change in Russia with respect to the energy sector. The Russian government seems to be developing a platform of political reform aimed at consolidating the energy resources in its own hands. Recent examples of this include the break-up of YUKOS, followed by the forced sale of its largest oil-producing subsidiary, Yuganskneftegas to an unknown shell company, Baikal Finance Group, and the subsequent sale of Yuganskneftegas to the state-owned oil company Rosneft. In addition, in the spring of 2005 there were plans for the Russian gas giant Gazprom (which is majority owned by the Russian government) to merge with Rosneft. However, this merger did not materialize as the government was not able to agree the terms. Finally, most recently, Gazprom acquired a 72.7% stake in Russian oil group Sibneft (previously owned by Milhouse Capital, which is generally understood to be owned by Roman Abramovitch and his business partners).

From a political perspective, it therefore comes as no particular surprise that, in comparison with the current Subsoil Law, the Draft Law focuses more specifically on the relationship between the State and the subsoil user. The Draft Law also envisages imposing significant restrictions on foreign investment in subsoil activities in Russia and reflects the government's intention to assert its control over Russian mineral deposits, which it considers to be of strategic importance.

This chapter will discuss these issues further and summarize the other key changes envisaged by the Draft Law.

Ownership of the subsoil

The Draft Law expressly states that the subsoil and mineral, energy and other resources contained therein are federal property.[3] It contemplates that all subsoil use rights for oil and gas deposits will be issued by the Federal Government (represented by the Ministry of Natural Resources of the Russian Federation).

Prior to August 2004, the issuance and revocation of subsoil licenses required the approval of both the federal government and the relevant regional authorities. This was known as the "two key" system and was in line with the Russian Constitution, which provides that natural resources (including the subsoil) are within the joint jurisdiction of the federal and regional authorities.[4]

Following amendments to the Subsoil Law in August 2004, however, the requirement for regional authorities' approval was removed in most instances. Currently the federal government alone issues and revokes most subsoil licenses.

Therefore, the Draft Law does not introduce any significant changes to the present regime. By stating that the subsoil is federal property, it continues the reform of federal consolidation and brings the legislation more in line with the existing system for the issuance of subsoil use rights.

Subsoil use rights

The Draft Law provides that subsoil use rights should be granted through an auction process. Detailed provisions are set out governing the organization and holding of such auctions.[5] Subsoil use rights may only be granted without an auction in specific cases envisaged by the Draft Law. Accordingly, the Draft Law abandons the notion of conducting tenders for subsoil use rights. This is viewed as a positive change as auctions allow for more transparent decision making as opposed to the tenders, which do not have concrete requisites and therefore could easily be subject to corruption.

The Draft Law lists the following types of subsoil use:[6]

- regional geological exploration;
- geological exploration;
- development and production of mineral deposits;
- extraction of underground water;
- construction and operation of underground facilities;
- special geological study for the purposes of construction of residential and industrial buildings;
- burial of radioactive, toxic and other hazardous waste; and
- gathering of mineralogical, paleontological and other geological collection materials.

Subsoil contracts

The Draft Law proposes to move away from the current licensing regime that regulates subsoil use, to a contractual regime that would enable the parties to negotiate the terms of a subsoil use contract. However, this freedom of contract is curtailed by the application of certain mandatory terms and the terms and conditions of the auction.

The Draft Law expressly provides for the possibility of entering into contracts for combined use, in relation to the rights to both geological exploration and development and production.[7]

The Draft Law provides that the parties to a subsoil use contract may negotiate the provisions of the contract (unless the terms are set by federal laws, other applicable law or by the terms and conditions of the auction).[8] However, the extent to which subsoil users may be able to negotiate terms is further limited in several respects:

1. the Draft Law provides that the authorized state body for subsoil use may develop model terms for subsoil use contracts depending on the type of use and the different mineral resources;
2. subsoil use contracts must contain certain mandatory terms; and
3. the Draft Law sets out strict (and short) deadlines within which the subsoil use contract must be entered into.[9]

The Draft Law envisages that a subsoil user, who has discovered a deposit in the course of geological exploration, will then have the right to conclude a contract for the development and production of that deposit. This is achieved by the Draft Law obliging the authorized state body[10] to enter into such a contract. If the authorized state entity refuses to enter into the contract, it may be challenged by the subsoil user in court.

Transfer of subsoil use rights

Subject to certain express restrictions, the Draft Law provides for the transfer and/or assignment of subsoil use rights insofar as the Draft Law and other federal laws permit.[11] The key restrictions are:

1. the rights may only be transferred/assigned to an entity that meets the requirements set by federal laws or by the auction organizers to subsoil users; and
2. the consent of the authorized state body with which the subsoil use contract was concluded must be obtained.[12] The authorized state body may, however, refuse to consent to the transfer. Such refusal may be challenged by the subsoil user in court.

Unlike the current Subsoil Law, the Draft Law permits the pledging of subsoil use rights.[13] Furthermore, and importantly, a pledge holder is not required to meet the requirements set by federal laws or the auction terms respectively.

The Draft Law provides that a subsoil use contract may be amended with the agreement of the parties to the contract.[14] In cases where agreement cannot be reached, the interested party may refer the case to the relevant court, which has the right to amend the subsoil use contract in situations envisaged by the Draft Law.

Termination of contract

According to the Draft Law, a subsoil use contract may be terminated with the agreement of its parties or, in certain cases envisaged by the Draft Law, upon a court ruling instigated by one of the parties.[15] A subsoil use contract may be terminated by a court following a petition of the government in the following circumstances:[16]

- if the subsoil user ceases to comply with a requirement set by the Draft Law, other applicable federal laws or a term or condition of the auction;
- if, through the fault of the subsoil user, deadlines for developing the deposit have been violated; and
- if the subsoil user fails to pay the one-time payment for subsoil use (i.e. the amount bid in the auction).

From a positive perspective, the Draft Law introduces greater clarity in relation to the procedure for amending and terminating a subsoil use contract. Once the subsoil use contract has been concluded it may only be amended or terminated by agreement of both parties or on the petition of one of the parties to the court. The Draft Law sets out the specific grounds on which the court may rely when making a decision to terminate or amend a subsoil use contract on the petition of the parties.[17]

The Draft Law sets out a list of violations that will not result in the termination of a subsoil use contract, but may result in a fine pursuant to a court ruling. These violations include:[18]

- failure to meet deadlines for the preparation of technical/project documentation;
- failure to provide information on the result of works;
- failure to meet deadlines for works;
- failure to meet deadlines to provide information on the amount of reserves;
- failure to meet deadlines for the execution of conservation works; and
- exceeding the maximum set amounts for the burial of industrial, radioactive, toxic and other dangerous waste.

The Draft Law introduces greater clarity with respect to the termination of subsoil use contracts and subsoil use rights. It sets out the circumstances in which the state may limit or terminate subsoil use rights (generally, where there is a threat to life and health, in the event of an emergency, natural disaster, military activities or other events of similar nature).[19] At the same time, it expressly gives the subsoil user the right to challenge in court the decision to limit or terminate his subsoil use rights.[20]

Restrictions on foreign participation

Under current legislation, there are relatively few restrictions on foreign companies or individuals with respect to holding subsoil rights. (While in practice, there are few examples of foreign companies holding a license direct, there are many instances of foreign entities holding licenses indirectly.) This position looks set to change significantly pursuant to the Draft Law.

According to the Draft Law, only Russian citizens or legal entities duly incorporated under the laws of the Russian Federation may be subsoil users.[21] Furthermore, the Draft Law gives organizers of auctions for subsoil rights the right to introduce restrictions that would prevent the participation in such auctions of Russian companies "controlled" by foreign entities.[22]

"Control" by foreign entities will be deemed to exist in respect of a Russian company if:

- (a) foreign individuals, individuals without citizenship and (or) foreign entities have the right to appoint the sole executive body and (or) more than 50% of the collegial executive body; and/or (b) where over 50% of the members of the board of directors (supervisory council) or another collegial management body have been elected on the recommendation of foreign individuals, individuals without citizenship and (or) foreign legal entities;
- the aggregate share of foreign individuals, individuals without citizenship and (or) foreign legal entities in the charter or share capital exceeds 50%;
- foreign individuals, individuals without citizenship and (or) foreign legal entities have the right to directly or indirectly control more than 50% of voting shares; and
- foreign individuals, individuals without citizenship and (or) foreign legal entities have the right to directly or indirectly control more than 50% of votes pertaining to the charter or share capital.[23]

These restrictions will not apply if the foreign entities "controlling" a Russian legal entity that would be a subsoil user, are themselves controlled by a Russian individual or a Russian legal entity.

The introduction of restrictions on the participation of Russian companies controlled by foreign entities in auctions will be a significant change to the current regime. Pursuant to the current Subsoil Law, auctions and tenders for the use of subsoil are organized by a federal executive body responsible for the management of the state subsoil fund.[24] The relevant body is currently the Federal Agency for Subsoil Use under the Ministry of Natural Resources (the "Agency"). The Agency, or its regional branch, takes decisions on the composition, procedure and terms of the tender/auction.[25] If the Draft Law is adopted in its current form, it will grant significant discretion to the auction organizers themselves to set terms and conditions, which could restrict the participation of Russian companies controlled by foreign entities.[26]

The restrictions may be manifested in the form of general restrictions in auction terms excluding foreign-controlled companies from participating in that auction. We understand that the restrictions are principally intended to apply with respect to auctions for the rights to develop large fields. The government's stated objective for the restrictions on foreign participation is to ensure state defense and security, including economic security.[27]

Strategic deposits

The restrictions on foreign participation in the Draft Law should also be read alongside the potential impact of an address given by President Vladimir Putin to the Federal Assembly on May 11, 2005. In this address, he stated that "*it is time we clearly determined the economic sectors where the interests of bolstering Russia's independence and security call for predominant control by national, including state, capital.*" One such economic sector is "*mineral deposits of strategic importance for the future of the country and future generations.*"

President Putin called for such decisions to be formalized in legislation and gave instructions to the government to prepare and submit to the State Duma draft federal laws restricting participation of foreign investment in Russia in certain sectors, including mineral deposits of strategic importance. The deadline for this was November 1, 2005. At the time of writing, the draft federal laws had not been publicly published.

However, the most recent statements of the Ministry of Natural Resources indicate how strategic deposits may be defined, essentially in three categories:

> The first category involves deposits of natural resources which are scarce in Russia – such as diamonds, uranium, high quality quartz and other rare minerals ... The second criterion involves a reserves size. Fields holding more than 150 million metric tons (1 billion bbl) of oil or 1 Tcm of gas will belong to the "strategic" group. These limits do not, however, cover Russia's continental shelf ... [there will be] a separate set of restrictions for offshore projects. The third factor concerns location, notably reserves situated on or close to sites associated with Russia's military bases, which could involve state security interests.[28]

In the meantime, the Draft Law already provides for the possibility of the government drawing up a list of "strategic" fields, for the development of which a special decision of the government will be required.[29] It is intended that the government will be able to draw up a list of strategic mineral deposits, in relation to which it will be possible to establish, by federal law, priority acquisition rights of the state.[30]

Current licenses

A final point to note is that the Draft Law provides that existing subsoil licenses will remain in effect upon their current terms. The subsoil users will not be required to obtain new subsoil licenses or to convert the existing subsoil licenses into contracts.[31] However, current subsoil users that are Russian legal entities will be entitled to convert their subsoil licenses into contracts at their discretion.

Draft amendments to the current Subsoil Law

It will inevitably take some time for the Draft Law to be approved and take effect in full. It is therefore likely that amendments to the subsoil use regime in Russia could be approved in the interim. The issue of foreign participation in subsoil projects is likely to be high on the priority list. Indeed, it has been reported that during the week of July 18, 2005, draft amendments to the current Subsoil Law, dealing with the issue of foreign participation, were submitted to the State Duma;[32] however, there does not seem to have been any significant progress with respect to such amendments. The amendments propose to restrict the participation of Russian companies controlled by foreigners in auctions for certain fields that have a special significance for the defense and economic security of the State.

Draft interim amendments may also introduce restrictions in respect of particular auctions or tenders, such that only Russian legal entities (or individuals) may participate in the auction/tender. These restrictions may also apply to the issuance of a production license based on a commercial discovery under an existing exploration license. The review of the Draft Law in the State Duma is likely to drive further developments in the near future.

Conclusion

As currently drafted, the Draft Law contains both positive and negative provisions and political messages. The positive aspects largely relate to technical and legal matters where, it is hoped, the prescriptive provisions will provide more clarity to the regime governing the issuance and operation of subsoil use licenses.

In substance, however, it is clear that the Draft Law is driven by strong political and policy considerations. These reflect the government's desire to regain its hold on the Russian energy sector. To this end, the Draft Law is unlikely to be completely welcomed by foreign investors and energy companies. Their ability to participate in Russian upstream projects is likely to be impeded by both the general restrictions on foreign participation and the specific restrictions in relation to "deposits of strategic importance."

Notes

1. Law No. 2395–1 "On Subsoil," dated February 21, 1992.
2. The government submitted to the State Duma Draft Law No. 187513–4 "On Subsoil" (the "Draft Law") on June 17, 2005. The Draft Law will enter into force once it is reviewed by the State Duma in three readings, approved by the Federation Council and signed by the President.
3. Article 10 section 2 of the Draft Law.
4. Article 72 of the Constitution of the Russian Federation adopted on December 12, 1993.
5. Article 36 sections 2 and 3 of the Draft Law.
6. Article 25 of the Draft Law.
7. Article 26 of the Draft Law.
8. Article 49 of the Draft Law.
9. Ibid. Pursuant to the Draft Law the state and the subsoil user should conclude the subsoil use contract within 30 days after the auction.
10. This was yet to be determined at the time of writing in October 2005. At that time subsoil use licenses were issued by the Federal Agency for Subsoil Use. Section 5.3.8 of the Regulations on the Federal Agency for Subsoil

Use (approved by Resolution No. 293 of the Government of the Russian Federation, dated June 17, 2004).

11. Article 54 of the Draft Law.
12. Article 54 section 1 of the Draft Law.
13. Article 55 section 1 of the Draft Law.
14. Article 78 section 1 of the Draft Law.
15. Article 79 section 1 of the Draft Law.
16. Article 79 section 3 of the Draft Law.
17. Articles 79 and 80 of the Draft Law.
18. Article 57 of the Draft Law.
19. Article 97 section 1 of the Draft Law.
20. Articles 95 and 97 of the Draft Law.
21. Article 9 section 1 of the Draft Law.
22. Article 60 section 5 of the Draft Law.
23. Ibid.
24. Article 13.1 of the Subsoil Law.
25. Ibid.
26. However, the Draft Law expressly prohibits the setting of such restrictions with the aim of restricting the freedom of economic activity and creating unjustified advantages for specific companies/individuals; if such a situation arises the restrictions may be challenged in court (Article 60 section 6 of the Draft Law).
27. Article 60 section 5 of the Draft Law.
28. *Nefte Compass*, October 20, 2005.
29. Article 23 section 3 of the Draft Law.
30. Article 23 section 5 of the Draft Law.
31. Article 121 section 2 of the Draft Law.
32. Overview of Russian Oil and Gas Legislation, September 2005.

6
Reserves Replacement in a Growing Demand Scenario: A Major Challenge to the Oil and Gas Industry

Álvaro F. Racero

After several years of research, in 1781 the Scotsman James Watt produced a rotary-motion steam engine, the engine of the industrial revolution that started to change the structure of an agriculturally based society towards an industrialized one.

In 1876, the German Nikolaus August Otto built the first four-stroke piston cycle internal combustion engine. The "Otto Cycle" was the first practical alternative to the steam engine and was the prototype of the combustion engines that have since been built. Since then, and as a consequence of Otto's invention, transportation has changed dramatically in its concepts and applications.

Towards the end of the nineteenth century, Anthony Lucas, an Austrian-born mining engineer, acquired the rights to drill an oil well in Beaumont, Texas. The well, known as Spindletop, used a new rotary technique. It was initiated on October 27, 1900, and on January 10, 1901 it blew-out at an estimated rate over 80,000 barrels of oil per day. Before this, a big producer flowed 50 barrels (8 cubic meters) per day. Lucas's well produced 1,600 times that amount. It showed that buried layers of rock could contain tremendous amounts of oil. What is more, it proved that rotary drilling was an effective way to tap that resource. Spindletop marked the beginning of the modern petroleum industry.

The twentieth century was the "hydrocarbon century." During this period, hydrocarbons fueled an industrialized society to standards of living unimaginable just a few generations earlier. The correlation between per capita energy consumption and development is a reflection of the social and economic structure of the world at the beginning of the twenty-first century.

The strong and persistent demand for oil and gas is a consequence of two major factors. The first is the increased consumption in developing countries, linked to a significant increase in world population, which has grown by 50% (from 4 to 6 billion) in the last 25 years. The second is the failure to develop acceptable alternative sources of energy.

The expected future demand for hydrocarbons varies from analyst to analyst; the EIA (US Energy Information Administration) projects an oil consumption of 120 million barrels per day in 2025. This represents an increase in production of 56% over the present level, a 2% per year rate of growth.

Can the oil and gas industry cope with this demand? Can reserves be found and economically developed to offset the declining fields and meet the increasing demand? And, in this process, what will be the equilibrium price?

There are two main elements that contribute to oil reserves replacement: "new oil" or oil that is accessed via exploration; and "old oil," which comes from discovered fields where the recovery efficiency is improved. Improved recovery efficiency is a function of both price and technological developments, which are also interlinked. Exploration alone has not been able to replace produced oil volumes since the early 1990s; hence worldwide oil reserves, in a broad sense, are currently decreasing.

The oil and gas industry will require significant capital, creativity and resources to overcome the giant challenge of providing energy over the next 25 years at the required pace and at affordable prices.

Introduction

This chapter addresses issues that are of importance to all concerned with the future of energy, because of their implications on the shape of the future of the world. These issues include oil and gas reserves, the world's energy outlook, its impact on the balance of supply and demand, and naturally, behind all this, the future prices of petroleum and gas. Given that oil and oil reserves are of higher concern today than gas, I shall give more emphasis to oil. Nevertheless, many of the concepts and ideas dealt with in this chapter are applicable to the current and future gas situation.

Like every forecast or projection about the future, this chapter represents a personal opinion, which will obviously differ from the opinions of many others. It is to be stressed that the content of this chapter and opinions expressed in it are the point of view of the author and do not represent an official perspective of his current employer.

This chapter intends to provide the reader with the basic elements that will shape future activity and price scenarios for the oil and gas industry. The latter is the eternal and fundamental question for all of us in the industry. I believe that no one has had nor will ever have the 100% correct answer.

Understanding reserves and resources

Projections of future supplies of oil and gas are measured in terms of "reserves" – that is, the oil and gas volumes that are or can be made available for consumption. The main role of the upstream sector in the industry is the transformation of "resources" to producible reserves, which can subsequently be commercialized.

Whenever an oil or gas field is discovered there is considerable uncertainty about the volumes that can be produced economically, i.e. the field's reserves. This uncertainty is reduced as production progresses. Nevertheless it remains – to a greater or lesser extent, depending on geological peculiarities and complexities – throughout the life of the field. Hence, reserves are better described by a probability distribution than by a single number or range of numbers.

The generally accepted worldwide definition of reserves and resources divides them into three categories:

- *Reserves* are those volumes of petroleum or gas that are known to exist and are susceptible to production under current economic conditions, hence, a function of price and technology.
- *Contingent resources* are those that have been discovered but are not viable for production under economic conditions and are being held until a breakthrough in technology or new developments or better economic conditions become available.
- *Prospective resources or Yet To Find (YTF)* mean petroleum or gas that has not been discovered but that the industry hopes to discover under normal exploration conditions.

Each of these can be further divided into three subcategories that correspond to the likelihood of the oil or gas actually being found and exploited.

For instance, in the case of reserves, proven reserves (1P) are those where it is 90% probable that the field will ultimately produce a volume equal to or greater than the volume calculated.

Proven + probable (2P) reserves are those with a 50% probability of being equal to or greater than the estimated volumes.

Proven + probable + possible (3P) reserves have only a 10% chance of being equal to or greater than the estimated volumes.

Figure 6.1 represents a typical reserves distribution curve versus probability for a field or a petroleum basin. The P50 (2P) defines the reserves that are most likely to be recovered by the end of the production life of a field. These are the reserves that industry typically uses to define the field development plan for new discoveries and, from a statistical viewpoint, are the ones to be used in forecasting field performance and future production.

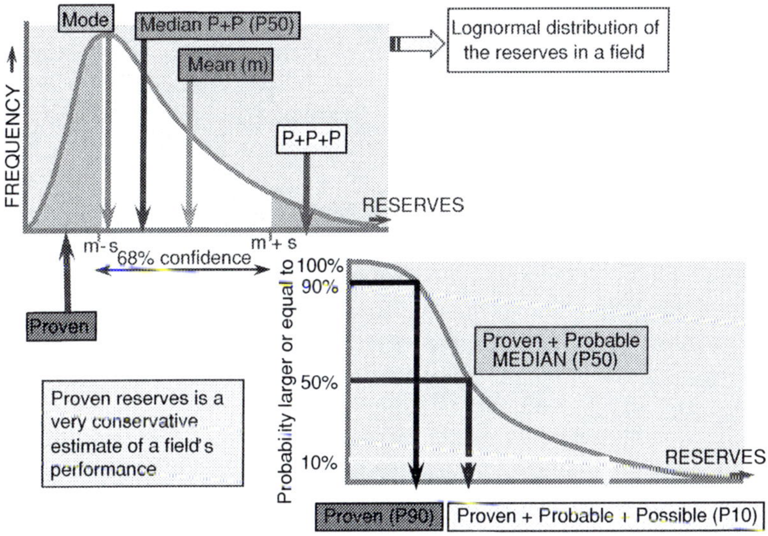

Source: Repsol YPF

Figure 6.1 Reserves Categories – Proven, Probable and Possible

Upstream corporations listed and traded on the New York Stock Exchange are obliged to report their proven reserves yearly. These proven reserves are to be defined according to the Securities and Exchange Commission (SEC) criteria.

It is important to stress that reserves are defined by the SEC as "hydrocarbons to be produced by a company" – which is different to the hydrocarbons to be produced by a field. For example, a gas accumulation is not allowed to book reserves unless there is a sales contract for the gas; until then, according to the SEC criteria, they are not considered reserves but contingent resources. To be classed as reserves, hydrocarbons have to be exploitable within the contract time frame and with today's proven technology; they cannot be booked based on new or emerging technology that has not been sufficiently proven.

A common mistake in forecasts is to consider the reported (SEC) reserves as indicative of a field's performance. There can be a vast difference between what a company understands as their reserves and what the SEC allows to be reported as such.

In summary, SEC reserves are a sort of financial indicator for oil and gas corporations, which in many cases does not reflect actual field performance. Therefore, SEC reserves (as proven reserves in general are at the initial stages of the development of a field) should not be used as a forecasting tool, since this would result in a fairly pessimistic, sometimes apocalyptic, view of reality.

Difficulties arise in forecasting because the P50 reserves are not systematically reported and are audited only internally. Confidentiality issues restrict the public sharing of these figures.

The remaining proven reserves of a field generally grow significantly during the appraisal, development and initial production phases of a field's life. This is commonly known as "reserves growth" or "reserves growth potential." Subsequently, remaining reserves decrease with production, resulting in an ultimate recovery similar to the 2P or P50 initial reserves estimates. This theoretical model is shown schematically in Figure 6.2.

Empirical studies have determined that the ultimate recovery of reserves continues to grow during the life of a field.[1] This is shown in Figure 6.3, for mature and well-known basins in the USA, Canada and the North Sea. Reserves growth approaches stable values after a period of 12 years in the Canadian example, but after more than 70 in the US case. This much longer time frame has to do with the peculiarities of the US hydrocarbon laws, which strongly promote the "strippers," or marginally economic wells, late in their productive life. In this case, the reserve growth is a reflection of the continuous activity in the onshore fields.

The reasons for this phenomenon are several: field extensions, deeper pools, oil in place unaccounted for (i.e. within silts and shale), improved oil recovery techniques, flexibility to take advantage of high price periods, better local knowledge of the resource, continuous reduction of operating costs, and so on.

By far the most difficult elements to estimate in a forecast or projection are the prospective resources or Yet To Find (YTF) volumes, given their

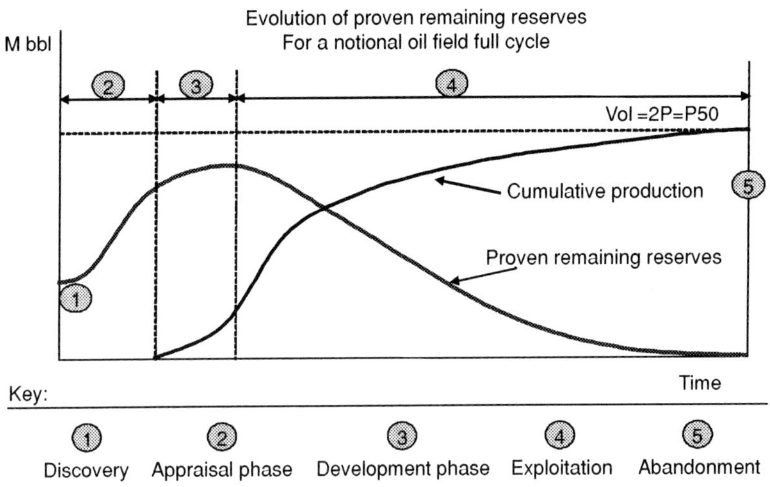

Source: Repsol YPF

Figure 6.2 Proven Remaining Reserves vs Time

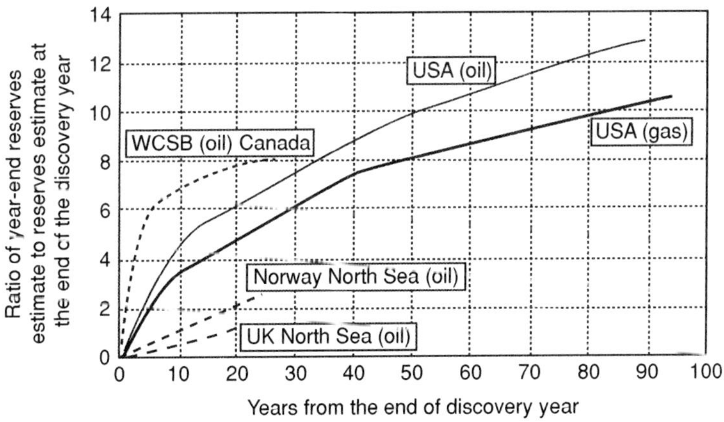

Source: Adapted from Baker, Wong and Virués (see Note 1 at end of this chapter)

Figure 6.3 Empirical Reserves Growth Estimates

intrinsic undefined nature. As an example, Figure 6.4 shows "the creaming curve method," which is a coarse estimate of the volumes YTF, applied to the US sector of the deepwater Gulf of Mexico.[2] The bases of this estimate are the similarities in areas and basic geological elements (reservoir rock, source rock, timing, etc.) between the deepwater fields (more than 300 meters water depth) and the fields discovered on the continental shelf. The volumes of the continental shelf (shown by the solid curve) are well calibrated, as thousands of exploration and development wells have been drilled since the 1950s, and there is therefore enough data available to reduce the uncertainty (note that this method tries to approximate the P50 reserves). That is not the case with the deepwater sector, which has only a couple of hundred exploration/appraisal and development wells (the short thick curve). The ultimate recovery for the deepwater fields cannot be accurately pinpointed, since the majority of the reserves are still to be discovered; however, on the basis of the existing data, it could be argued that they should, at least, double the discovered reserves to date, surpassing the 30 billion barrels of oil equivalent (the dashed curve).

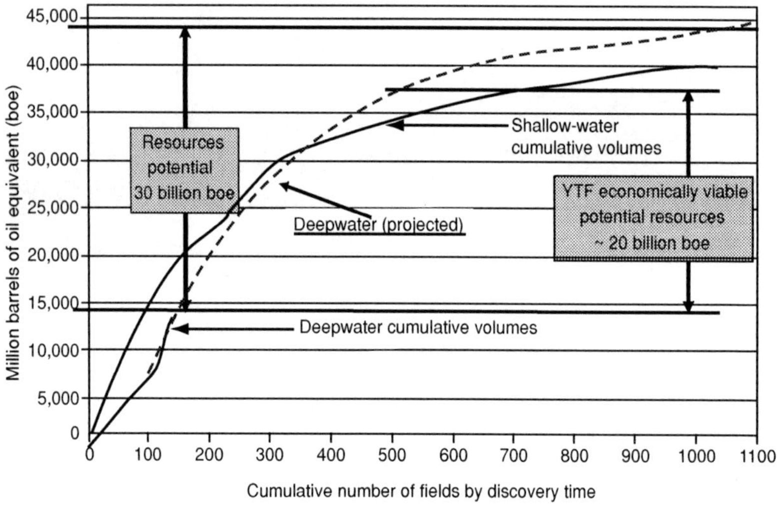

Modified from MMS 2003

Source: Repsol YPF

Figure 6.4 The Deepwater Gulf of Mexico, Estimate of YTF

The *Statistical Review of World Energy*, issued by BP,[3] is a magnificent compilation of the estimates of worldwide reserves. The June 2005 edition compiles the figures, as of December 31, 2004, of the remaining proven reserves. Figure 6.5 shows the remaining proven reserves for the last 25 years. As can be observed, these have been growing throughout the period; however, the growth rate is clearly decelerating.

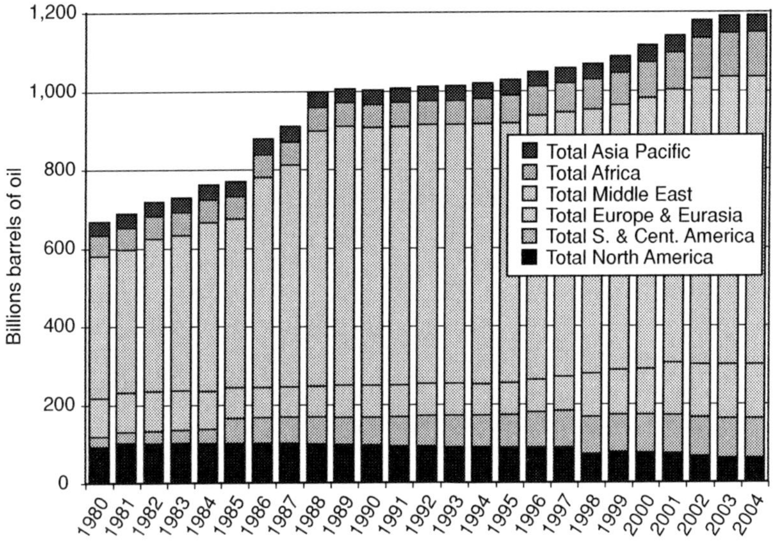

Source: Modified from BP

Figure 6.5 World Oil – Remaining Proven Reserves

As is well known, the majority of the worldwide reserves are concentrated in the Middle East, an area that is today geopolitically unstable. Two countries within the Middle East, Saudi Arabia and Iraq, combine to make up one-third of the worldwide reserves. This is the first element behind the high price of crude today: the oil is concentrated in countries whose geopolitical situation is unpredictable. Note in Figure 6.5 how little weight is carried by the other geographic zones. If we had to emphasize two that had been important, they would be the Former Soviet Union (FSU) and Africa, both areas of low consumption but notable reserves. However, nothing can compare to the concentration of reserves in the Middle East.

Given current rates of production, it is estimated that the remaining proven reserves will be exhausted in 40 years. However, the following points should be borne in mind when considering this estimate:

- The remaining proven reserves do not take into account the probable reserves that are likely to be incorporated to proven reserves in the appraisal and development phases, nor the empirically demonstrated reserves growth. This is of particular importance for fields and basins that are at early stages of appraisal and development (i.e. deepwater in general, the Caspian basin, etc.)

- Contingent resources are also not taken into consideration. Technology and higher prices will result in an increase of proven reserves from these current resources (i.e. heavy oil and tar sands).
- "Yet To Find" volumes are unaccounted for. This can be of particular relevance for some of the major resource-holding countries that have been closed to international exploration investment for decades (i.e. Saudi Arabia, Iran, Iraq and Mexico) and unexplored or under-explored areas (ultra-deepwater, polar oil, overpressure pools, tectonically complex areas, etc.).
- On the other hand, the forecast increase in demand should result in a reduction of the reserves to production ratios.

The case of gas is much less mature than oil, with a remaining proven reserves-to-production forecast of some 67 years. Furthermore, the development of LNG (liquefied natural gas) plants and additional markets should result in significant movements from discovered contingent resources ("stranded gas") to proven reserves in the coming years.

As the above examples show, establishing a figure for the ultimate recovery of oil in the world is far from being a simple task and is open to controversy. Figure 6.6 plots several estimates of ultimate oil reserves

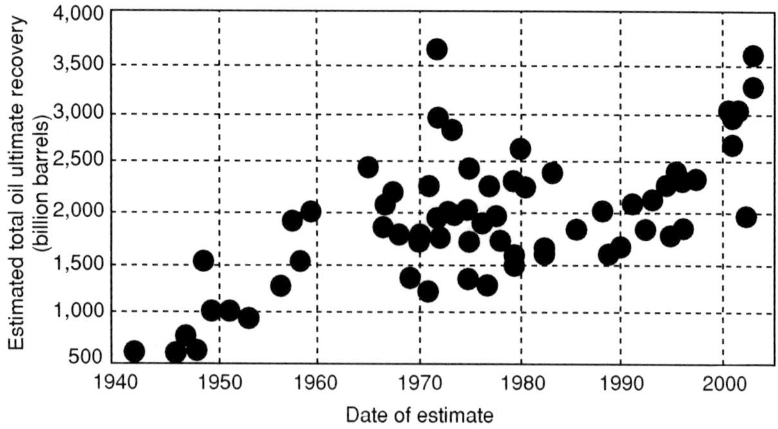

Source: Adapted from Lynch (see Note 4 at end of this chapter)

Figure 6.6 Estimates of Oil Ultimate Recovery

recovery through time – excluding heavy tar sands.[4] As can be observed, the dispersion is quite significant, but there is a general trend of increase in the estimates over time. This can be explained as new productive basins and technologies are incorporated. To put this graph into context, it is useful to bear in mind that: (1) approximately 950 billion barrels have been produced to date, and (2) current annual global consumption is some 30 billion barrels.

Demand drivers

The international and global characteristics of the oil industry are a reflection of the nature of supply and demand, geopolitical differences and geographic distances. Whereas supply is concentrated in the Arabian Gulf, the FSU, West Africa and some in North and South America, the majority of demand is located in the USA, Europe, India, China and Japan. In 2004, North America, Europe and Eurasia, and Asia Pacific accounted for 84% of total oil consumption, which was 29,314,938,600 barrels. The Middle East consumed a mere 7%.

Another challenge that the industry has been dealing with lies in the distances from production areas to demand centers, which sometimes cause a bottleneck in transportation. Furthermore, cycles in transportation costs complicate the view on final consumer prices. In addition to the transportation bottleneck, there is also one in refining, where the scenario is further complicated by the high price of oil and by refineries with significant deficits and stratospheric refining margins. In summary, the situation facing the final consumer is not an easy one; he is supporting high costs at origin, expensive and volatile transportation costs and high refining costs. Furthermore, in Europe and parts of South East Asia, a significant tax is also applied on petrol.

Despite the high prices, however, the world continues to experience a strong and persistent demand for oil and gas. This is a consequence of two major factors. The first is the increased consumption in developing countries, linked to a significant increase in world population, which has grown by 50% (from 4 to 6 billion) in the last 25 years. The second is the failure to develop accepted alternative sources of energy.

The expected future demand for hydrocarbons varies depending on analyst to analyst. The EIA (US Energy Information Administration) projects an oil consumption of 120 million barrels per day in 2025.[5] This requires an increase in production of 56% over the present level. Whether this level of production is achievable and sustainable is questioned by several analysts. But if the forecast is accurate, total production for the period 2005–25 would need to be 776 billion barrels.

Figure 6.7 shows the forecast total demand, which is about a 2% increase per year, and the demand of the main sectors. As can be observed, transportation is the main driver of the increase in demand, followed by industrial use of oil, which includes electric power generation. Geographically, the main driver is the emerging Asia region, which is forecast to more than double its consumption, from 16 million barrels per day (MMbpd) in 2004 to 34 MMbpd in 2025.

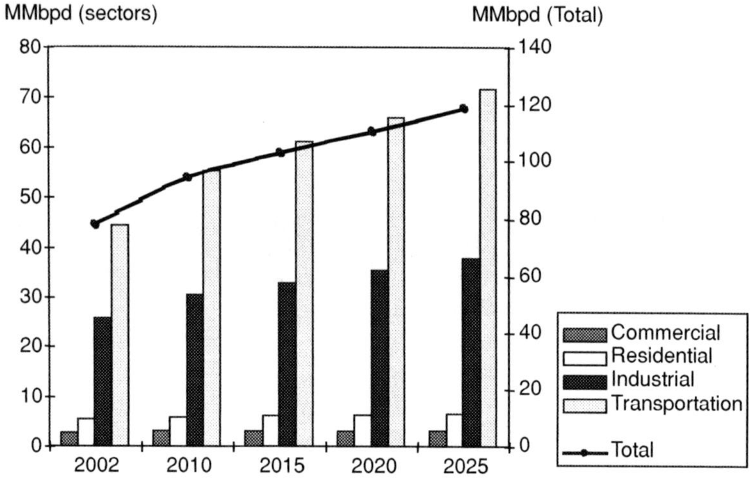

Source: Adapted from EIA

Figure 6.7 Oil Demand Forecast

From Figure 6.7 it can be concluded that alternatives to gasoline and increased fuel efficiency in transportation are essential elements for balancing supply and demand in the future. Unfortunately, it is in the transportation sector that most difficulties in introducing alternatives to oil are to be found.

As previously mentioned, the fundamental cause of the increase in demand is the growth of the world's population. This fact and its impact on energy demand are shown in Figure 6.8.[6]

The black solid line is the worldwide population expressed in billions. As can be observed, and as a consequence of fundamental achievements in the field of medicine, the steady but gentle growth rate from 1850 to

1950 breaks upwards from a composite average growth rate (CAGR) of 6 million people per year to a CAGR of 76 million people per year for the period 1950 to 2003. The fact is that the world population has doubled in the last 40 years; this has tremendous implications, among them a much larger active population, in particular in the developing world, that drives a lower cost industrial/manufacturing revolution with the corresponding demand for energy. South East Asia, China and India are prime examples of this phenomenon. The increased demand for energy manifests itself predominantly as an increased demand for oil.

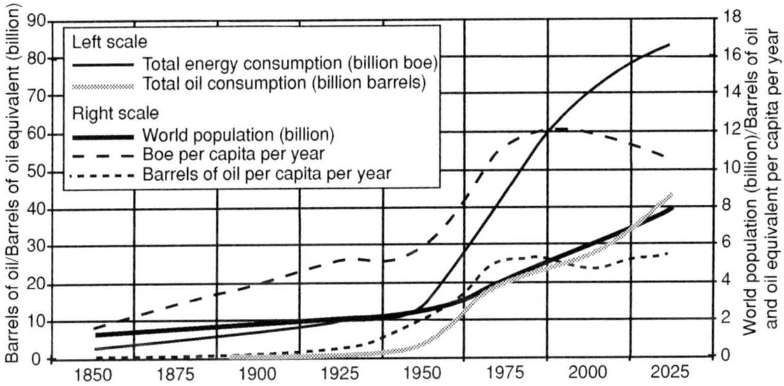

Source: Adapted from Laherrere (see Note 6 at end of this chapter)

Figure 6.8 World Population and Energy Consumption

The thin black solid line corresponds to total world energy consumption expressed in billions of barrels of oil equivalent. Oil accounts currently for 40% of the total energy consumption. In this case, a phenomenal growth for the period 1950–75 can be observed, a consequence of the previously mentioned growth of population and per capita energy consumption (represented by the upper dashed black line), which almost doubled during this period. The oil embargos related to the 1973 Yom Kippur war started to curb this trend, which then declined over the next 30 years. It is expected that energy (oil) efficiency, expressed as per capita consumption, will further improve significantly in the future, becoming a fundamental source of "new reserves" expressed as lower projected demand. There must also be increased efforts to improve energy efficiency, a key source of additional energy supply. Environmental pressure and the Kyoto Protocol should help in achieving this higher efficiency.

Factors determining future oil demand are well defined. Economic development in Asia will be crucial to long-term growth in oil markets. China, India, and the other emerging nations of Asia are expected to experience a combined economic growth of 5.5% per year between 2002 and 2025, according to the base-case of the EIA forecast, the highest rate of growth in the world. This robust expansion in gross domestic product (GDP) translates to a 3.5% annual increase in regional oil use. The projected upsurge in Asian oil demand could further strengthen economic ties between the Middle East and Asia, as Asian nations come to rely more and more on Middle East oil supplies.

Despite the phenomenal increase in population the world as a whole is getting wealthier and the economies of poor countries are developing, yet poverty remains a major problem. Nearly four in five people in the world are considered poor; existing on less than US$10 per day. The availability of affordable energy will be critical in raising their standard of living. Meanwhile, many of the world's ecosystems are under stress, and will become more so as an increased population puts more and more demand on resources.

Supply options

We have already indicated how oil supply is, and will be even more in the future, concentrated in the producing states of the Arabian Gulf. The Middle East region is currently producing 30% of the world's oil and accounts for almost two-thirds of the total proven oil reserves (Figure 6.9).

A very important effect of the increase of demand on supply is the reduction in spare capacity, which has shrunk almost to zero. Hence, the upstream sector has moved from being a buyer's to a seller's market, and new producers, such as the Caspian Sea, have emerged on the international scene.

Table 6.1 is taken from the EIA's *International Energy Outlook* (2005) and shows estimates of the conventional oil resource base by region. Proved (remaining) reserves are from the annual assessment of worldwide reserves published by *Oil & Gas Journal* (note that this figure of 1,277 billion barrels is similar to the BP estimate of 1,188 billion barrels shown in Figure 6.5). Reserves growth and undiscovered estimates are based on the World Petroleum Assessment 2004 by the US Geological Survey (USGS). The oil resource base is defined by the three categories discussed early in this chapter: proven remaining reserves; reserve growth; and undiscovered (YTF). Some analysts, however, find the EIA viewpoint

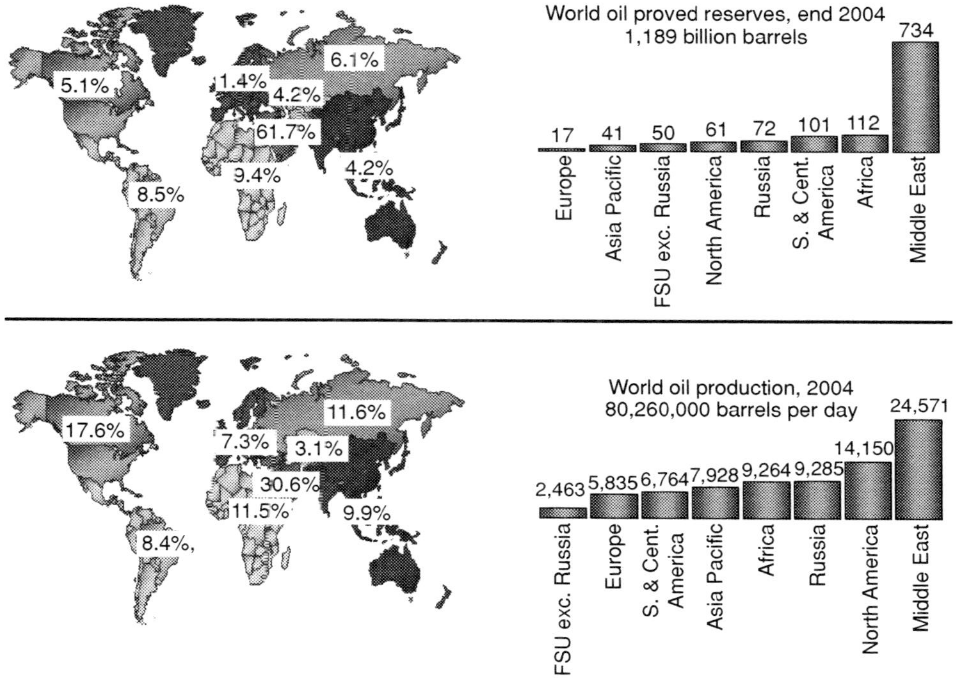

Source: Repsol YPF, based on BP and own data

Figure 6.9 Oil Proved Reserves and Production

overly optimistic, because of what they consider to be a significant overestimation of both proven reserves and ultimately recoverable resources. A reliable and comprehensive inventory of reserves, resources and YTF is in everyone's best interest.

YTF information in Table 6.1 is projected to the year 2025 (i.e. volumes to be found up to that date) and is derived from the USGS mean estimate, which is an average assessment over a wide range of uncertainty for undiscovered resources; it represents 2P values (proven + probable reserves). Figures are in billions of barrels.

Table 6.1 World Conventional Oil Resource Base by Region

Region	Proved reserves	Reserves growth	YTF undiscovered	Total
Mature market economies				
United States	21.9	76.0	83.0	180.9
Canada	178.8	12.5	32.6	223.9
Mexico	14.6	25.6	45.8	86.0
Western Europe	15.8	19.3	34.6	69.7
Japan	0.1	0.1	0.3	0.5
Australia/New Zealand	1.5	2.7	5.9	10.1
Transitional economies				
Former Soviet Union	77.8	137.7	170.8	386.3
Eastern Europe	1.5	1.5	1.4	4.4
Emerging economies				
China	18.3	19.6	14.6	52.5
India	5.4	3.8	6.8	16.0
Other emerging Asia	11.0	14.6	23.9	49.5
Middle East	729.6	252.5	269.2	1,251.3
Africa	100.8	73.5	124.7	299.0
Cent. and S. America	100.6	90.8	125.3	316.7
Total World	**1,277.7**	**730.2**	**938.9**	**2,946.8**
OPEC	885.2	395.6	400.5	1,681.3
Non-OPEC	392.5	334.6	538.4	1,265.5

Source: EIA (figures are billions of barrels of oil)

We have already discussed that reserves growth is a reflection of increased efficiency in the exploitation of discovered reserves. The estimates presented in this table show the highest potential for those areas with the most significant oil-in-place (the Middle East and FSU) that, for various reasons, still have the potential to increase recovery efficiency.

Reserves replacement

If you had walked about in Denver airport in January 2006, as I did at the time of writing, you would have seen a Chevron advertisement stating that "The world consumes two barrels of oil for every barrel discovered." This statement neatly summarizes the current state of affairs: we are using up oil reserves more quickly than we are replacing them. And, as discussed in the previous sections, the strong increase in demand forecast for the next 20 years aggravates the problem.

There are two main elements that contribute to oil reserves replacement – "new oil," oil that is accessed via exploration; and "old oil," which corresponds to discovered fields where recovery efficiency is improved. Improved recovery efficiency is a function of both price and technological developments, which are also interlinked.

According to data compiled by Campbell and Zagar,[7] since the early 1990s exploration alone has been unable to replace produced oil volumes; hence net world oil reserves have been decreasing since then. This is illustrated in Figure 6.10, which shows the reserves (approximation to 2P) backdated to the year of discovery. The projection is somewhat pessimistic, since it cannot fully incorporate the "reserves growth" discussed earlier in this chapter. Nevertheless, it is apparent that the amount of "new" oil discovered each year has been decreasing since, at least, 1990. Although this may be counterintuitive, note that it is compatible with the increase of proven reserves shown in Figure 6.5, because of the "reserves growth" phenomenon.

To set Figure 6.10 in perspective, it is also relevant to realize that several countries with a high exploration potential have been closed to private investment during the 1970s, 1980s and 1990s. Over the past ten years the oil sector has experienced a restricted opening of a majority of the previously closed regions, including most of the OPEC countries and FSU giving IOCs (International Oil Companies) access, albeit limited, to a new playing field. Iraq, Saudi Arabia and Mexico remain largely closed to outside investors, although the last two have recently offered a cautious opening of the dry gas sector. Kazakhstan, with the discovery of Kashagan, which has over 10 billion barrels of oil of estimated 2P reserves, is an example of what additional reserves might be expected from more intensive exploration of these regions. If this happens, it will significantly increase the value of future discoveries over what is shown here.

Referring back to Table 6.1, the remaining exploration potential is indicated by the undiscovered YTF volumes column. These data are the mean values of the probability evaluations that the USGS carries out

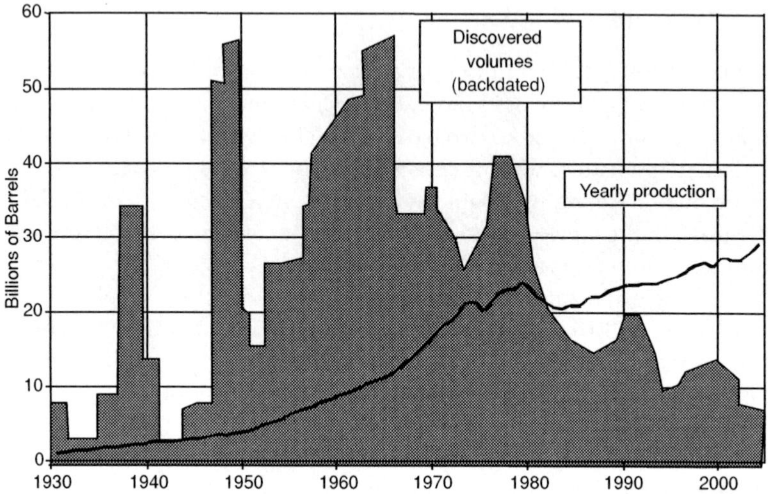

Source: Based on Campbell and Zagar (see Note 7 at end of this chapter)

Figure 6.10 Exploration Performance

on a continuous project basis. Results are regularly updated and made available through the organization's web site. A total of 940 billion barrels are estimated as a 2P mean value (reserves + reserves growth) to be found during the next 20 years. Examining this table we can depict the major contributors.

Despite its high degree of exploration maturity, the United States is expected to contribute 83 billion barrels, almost 10% of the total. These volumes are considered to be present in areas with potential, but of low exploration intensity, due either to moratoriums (Atlantic margin, eastern Gulf of Mexico and onshore Alaska) or to technical difficulties that are anticipated to be overcome in the forecast period (overpressured pools, ultra-deepwater, polar offshore Alaska).

Mexico is currently the fourth largest net oil exporter in the world and the first in the Americas. This positive balance coupled with limited recent exploration investment has resulted in a significant potential of undiscovered reserves in the deepwater offshore sector of the Gulf of Mexico. The figure of some 46 billion barrels estimated by the USGS is supported by other studies and evaluations.

The Former Soviet Union has, according to these evaluations, a YTF potential of some 170 billion barrels. A large part of this volume is allocated to the northern polar basins.

The Middle East has by far the largest potential, with a mean value estimated at some 270 billion barrels. The majority of this is contributed by Iran and Iraq, countries where there has been very limited exploration during the last 20 years because of military conflicts and the associated political instability.

Oil price and price projections will be discussed in a later section of this chapter, but we can anticipate that prices have had a great impact on the exploration effort through time. The 1990s were marked by declining prices in real terms, sliding down from a high during the first Iraq War in 1991 to a low in 1998, with Brent crude trading close to $10 per barrel. This downturn in prices resulted in a reduction in capital expenditure that affected mainly exploration activity, R&D funding, contracted services markets and infrastructure. Some of the root causes for the current situation, in particular issues related to production capacity, probably lie in the limited investments during the last decade of the "hydrocarbon century."

In the social arena, the second half of the 1990s witnessed the development of the internet and the onset of globalization. The internet has, in a subtle way, changed our day-to-day life, but it is probably in the investment and financial sectors that most changes have been experienced. The dynamics of these sectors have changed dramatically due to the accessibility to information, free movement of capital and the subsequent proliferation of investment funds and financial analysis. This market-driven focus on short-term profitability and related performance metrics during the low cycle commodity prices of the last decade did negatively impact exploration expenditures. Key objectives were then set to increase production, and to minimize costs and spend in order to maximize margins, revenue, cash flow and earnings. In this context, legacy assets provided the cash and funds for a series of mergers and acquisitions that resulted in a significant consolidation in the industry. Hence, for many companies acquisition and development of already discovered reserves became the operating strategy of choice for reserves replacement and production growth. Maxted et al.[8] tracked the evolution of yearly industry expenditure during this period. They show that while acquisition expenditure from 1998 to 2000 swelled and then receded, development expenditure was first flat and then grew. Meanwhile, exploration remains almost constant in the range of 10–20 billion US dollars per year. Although, majors have since been taken over by independents in terms of their share of spending on exploration, on the industry level, exploration spend as share of total expenditure has reduced progressively over time.

In relative terms, exploration spend has reduced progressively over the period for the majors, from a 25% of the total budget in 1994 to merely a 14% in 2004. Contrary to this, the independents have maintained a higher level of exploration expenditure, in the 25% to 35% range, for the same period. There is a clear transfer of exploration risk from majors to independents during this period. The result of this has been a significant reduction in people through industry consolidation during the past 15 years, including a substantial loss of explorers, no next generation, and evolution towards a risk conscious culture that has forced many explorers to engage in either acquisition/exploitation activity or corporate restructuring.

The challenges that the exploration function is facing are twofold. The first is regaining business confidence in the ability to be successful (particularly from corporate leaders and the market) while establishing a track record of successful exploration, which delivers value. The second is exploring for a resource that is more difficult to find, at a competitive cost, with fewer people.

The last three years have been characterized by a significant increase in oil price. The business models emerging from this new situation are not yet well defined and, somehow, a pattern of polarization of the existing ones has been observed. During 2005 we witnessed two major acquisitions (Unocal and Burlington) by major oil companies and aggressive exploration bidding by independents (Oxy in their comeback to Libya), while exploration budgets of the major oil companies experienced only a minor prudent increase. This is a concern for several analysts who believe that if oil and gas are to fuel economic growth in the twenty-first century, large operators must give higher priority to the replacement of their reserves through exploration rather than acquisition, which does not increase total reserves.

The fear of future shortages in the access to supply has forced the national companies of emerging Asia into the international arena. The fact that these new players are not ruled by the metrics of Wall Street creates an unprecedented type of competition in the marketplace. However, it is fair to say that securing the supply of hydrocarbons is critical to economic development for those nations.

A large majority of current production comes from mature fields that were discovered decades ago (see Figure 6.10) and are therefore now in decline. Major oil companies have quoted decline rates of the order of 10% for their currently operated fields. In Venezuela the rate is assumed to be between 15% and 25%. In Figure 6.11, a 5% yearly decline rate has been assumed for current production. This means that the group of existing reservoirs that are currently producing 80 million barrels

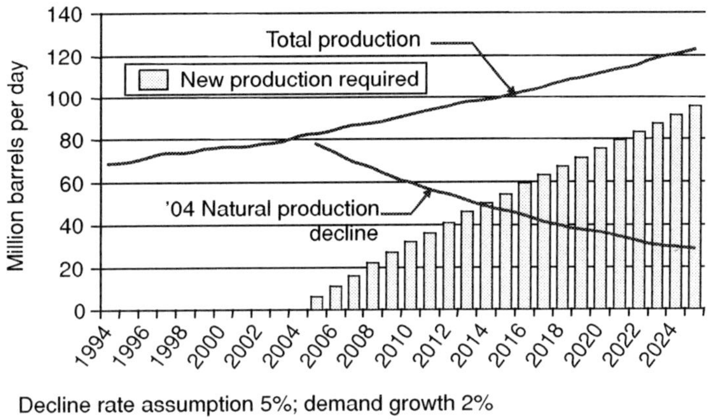

Source: Repsol YPF

Figure 6.11 Impact of Decline Rate

per day will have a combined production of some 30 MMbpd by 2025. Note that this natural decline is compatible with the reserves growth previously mentioned. Redevelopments of mature fields to increase their ultimate recovery are one of the main sources of additional reserves, but additional capital expenditure and new technologies are required to offset the natural decline of a reservoir. The challenge of "fighting the decline," a term used by the reservoir engineers, is a major one for the industry to overcome in the future, as more and more large fields enter a period of advanced maturity.

Following the 1986 oil counter-shock and the low prices that prevailed in the 1990s, the main priority for the industry was to reduce costs. Oil companies did streamline their operations, but also widely implemented new technologies such as 3D seismic and horizontal drilling. This resulted in a significant increase in the recovery factors and a dramatic decrease in costs in non-OPEC countries. Between 1986 and 2000 the average development cost of non-OPEC supplies was reduced by a factor of two.[9] During the last five years we have witnessed an almost twofold reduction in "conventional" deepwater development cost and start-ups (time from discovery to first oil) in the Gulf of Mexico.

We have already noted that the concept of reserves is an economic one – a function of price and costs. It is also widely accepted that the majority of the world's conventional oil that is accessible has been identified and accounted for in the forecasts. We can conclude that in the next 20 years we will be running low on cheap, high quality, conventional oil. Therefore the industry should turn its attention to the unconventional and the difficult resources that will take investment, technology, and innovative thinking. New oil supplies will be found in offshore deep water, the Arctic regions, tectonically complex zones, overpressured reservoirs (which are becoming more reachable thanks to the underbalanced drilling and expandable tubulars technologies) and in the development of heavy crude. Where costs were once prohibitive, new technologies enable energy companies to extract oil and gas from existing reservoirs. Offshore technologies make it possible to find and extract oil and gas in deep ocean water – a location that was considered inaccessible just a few years ago.

The resources of heavy oil are as large as 4,000 to 5,000 billion barrels in-place, of which the potential reserves can range between 500 to 1,000 billion barrels. The technological challenge resides in the fact that some 80% of it is extra-heavy oil (20°–7° API, hardly mobile at reservoir conditions) or tar sands and bitumen (12°–7° API, non-mobile at reservoir conditions). Geographically the heavy oil accumulations are largely concentrated in North and South America, mainly in Canada and Venezuela.

According to Merrill Lynch, development marginal costs for some conventional oil resources after contract range from Saudia Arabia with about $5 per barrel to marginal Canadian oil sands of around $45 per barrel.[10] A sustained price like the one averaged for 2005 (some $54 per barrel for Brent) will make a large number of unconventional oil resources economically viable, in particular the Canadian heavy oil and oil (tar) sands.

As a summary of this section, we can conclude that oil reserves replacement in the forecast demand period is not straightforward and will not be cheap. Figure 6.11 showed that as early as 2009, the industry will have to increase its production by 20 MMbpd of "new production" (output from new fields less natural decline of existing fields). That figure will double by 2013 and will reach a volume equal to the current production of 80 MMbpd by 2022.

In the exploration arena, "new oil," the large majority of significant discoveries are likely to fall into three main categories:

- Technologically demanding environments such as overpressured pools, tectonically complex areas, deep water, ultra-deep water and the Arctic.
- Opening of prospective new areas (Atlantic margin of the USA, Iran, Iraq, offshore Mexico, etc.).

- New concepts in known producing basins supported by superior geophysical tools (e.g. seismic depth imaging).

The reserves growth, the "old oil," will still account for the majority of "new production." The main driver will continue to be technology. Experience tells me that this is one of the fastest-moving and more creative technology-based industries in the world. The boundaries of knowledge in every aspect of the oil and gas industry are moving forward every day.

Increased efficiency in field development and production has to be coupled with much needed changes in infrastructure, which will be translated into increased production. The relatively low commodity price of the 1990s has left a legacy of low capital investment, which in turn has left several areas with limited short-term production growth due to lack of infrastructure. The EIA estimates that an average capital investment of US$560 billion per year, during the next 20 years, is required to meet projected demand.

The factors that will ensure the required significant contribution from the existing fields are threefold:

- Continuous cost reduction for each technology, prolonging the commercial life of old fields and optimizing development of young ones.
- Increased recovery efficiency through better and deeper understanding of the reservoirs.
- Economic viability of massive and well defined heavy oil resources.

The areas where a more significant contribution is expected are those with higher in-place volumes, namely the Middle East, FSU and the heavy oil of Venezuela and Canada. Neither "old" nor "new" oil has a clear low-cost scenario. Exploring in challenging frontier areas and increasing recovery efficiency in mature fields or heavy crude are all expensive exercises.

Oil prices

Several factors have influenced and determined the evolution of oil prices in the past. Many of these same factors will determine what happens to prices in the future.

Structural changes in price can be observed in the late nineteenth century, a consequence of oversupply as Russian and other international oil started entering the marketplace. The period 1900 to 1973 saw a steady overall decline in prices in real terms, caused by an industry finding more and more oil and becoming more and more efficient in areas such as production, transportation and refining. The main structural change in oil prices in recent history occurred during the 1970s. Sparked by geopolitical differences, it was sustained by the consolidation of OPEC into an effective cartel.

The significant increases in price during the last two years, coupled with the strong forecast demand, have raised the issue of a possible structural change in the market, which would result in high prices in the future. Given the impact that oil prices have in conditioning the performance of the world economy, a genuine concern is generating intense debate about this matter.

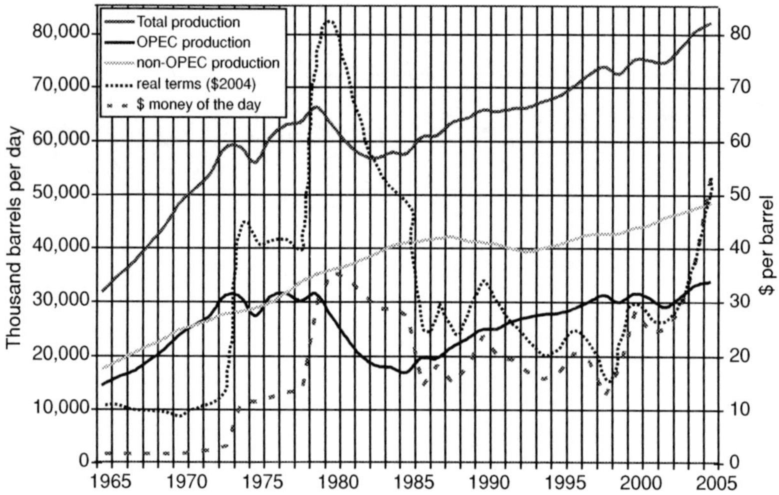

Source: Repsol YPF

Figure 6.12 Oil Prices – From the Oil Crisis to Today

Figure 6.12 shows oil price evolution and OPEC and non-OPEC production over the last 40 years. This figure can help us to analyze the structural price change in the 1970s and the current situation with supply options, knowing that at the moment we have a clearly demand-driven market.

In the period 1965–73 total demand almost doubled (as is shown by the "total production" line). This demand is mainly satisfied by an increase in OPEC production, which also doubled during the same period. Non-OPEC production grew at a more modest pace; in 1973 OPEC produced more than 50% of the world's total oil. The political crisis in the Middle East during the 1970s led to a reduction of supply by several of the main producers in the region, sending the price of a barrel shooting up sevenfold in real terms and more than tenfold in the money of the day (dashed lines). The first half of the 1980s therefore saw a reduction in demand, with many of the world's large economies entering a period of stagnation or recession. OPEC cut production to shore up a rapidly falling price, and in 1985 its market share hit its lowest level, less than 30%.

In the second half of the 1980s demand started to grow again, balancing the price in a couple of cycles. Non-OPEC production remained flat at marginal costs. The increase was met by OPEC producers following a complex quota formula that by the early 1990s had created significant confrontations among its members.

A period of economic recovery during the 1990s brought a further increase in demand. 1979 production levels were reached again in 1993, although OPEC's contribution was only 41%, as against 47% in 1979. A decade of internal disagreements within OPEC had resulted in a supply-dominated market with downward-sliding prices, which helped the general economic recovery. For the second half of the 1990s the increase in production from the non-OPEC sector was due mainly to the contribution of the Russian Federation. The years following the price rebound (1999–2002) can be seen as a transitional period to the present situation.

It is a fact that since 2000 OPEC has become a more effective organization than it had been in the previous decade. Furthermore, the previously mentioned increase in demand has resulted in greater tension in the market and put pressure on the infrastructure. This situation was accentuated during 2005 by natural disasters in the Gulf of Mexico and a growing uncertainty over the political future of Iran and Iraq.

Oil prices have been highly volatile over the past 25 years, and periods of price volatility can be expected in the future, principally because of unpredictable political and economic circumstances. It is widely recognized that tensions in the Middle East, for example, could give rise to serious disruptions in normal oil production and trading patterns.

Historically, probably the only truth common to all oil price forecasts is that they have been proven wrong. With a low probability of success, therefore, I prefer to opt for prudence. Nevertheless, we can try to highlight some of the factors that will have a strong influence on future oil prices.

The first one is demand. We have seen how in the past a significant price increase resulted in a reduction in demand that eventually pulled down the price. Some analysts believe that the current situation is different and that the present demand is stronger than ever before. If we observe the aggressive international positioning of the emerging Asian national oil companies, it invites us to believe that future demand and prices will be high. On the other hand, the more prudent behavior of the Western oil companies balances that perception.

The supply side seems a bit clearer. We saw in Figure 6.12 that the significant increase in demand over the last four years has been covered mainly by OPEC production. This is proof that non-OPEC producers (with the possible exception of the FSU) simply do not have the capacity to increase production significantly in the short term. Furthermore, as has previously been discussed, the marginal cost of maintaining (or increasing) production for non-OPEC producers is likely to be high.

It is to be expected then that OPEC, and the Middle East in particular, will have to absorb a large majority of the future oil demand. It is impossible for me even to try to formulate hypotheses on how the complex future of the Middle East is going to unfold, but it is clear that this is the single most important factor in determining the future price of oil.

Conclusion and the oil ultimate recovery debate

As with every period of high prices the logical concern about scarcity appears on stage. However, the problems facing us at the moment are not about scarcity; they are connected with lack of capacity and the difficulties of increasing that capacity to keep pace with increasing demand. As we have seen, the oil crisis happened because OPEC producers were able to cut back production and squeeze up prices, but there was never any indication of a scarcity as such.

Figure 6.13 summarizes the statistics that have been discussed in this chapter and puts them into context. The low-end and high-end estimates for ultimate oil recoveries are the ones previously shown in Figure 6.6. A range of development costs has been included in the Y-axis for the expected average type of resources of each of the trenches. These costs are pure project costs (in 2005 dollars) and assume no cost reduction due to technology. With the assumption of a constant daily production of 100 million barrels (which is the estimated average for the period 2005–25), a tentative time line is also plotted. The debate about ultimate oil recovery has been, is, and always will be intense, passionate and controversial.[11] During the last five years I have had the opportunity to live on three

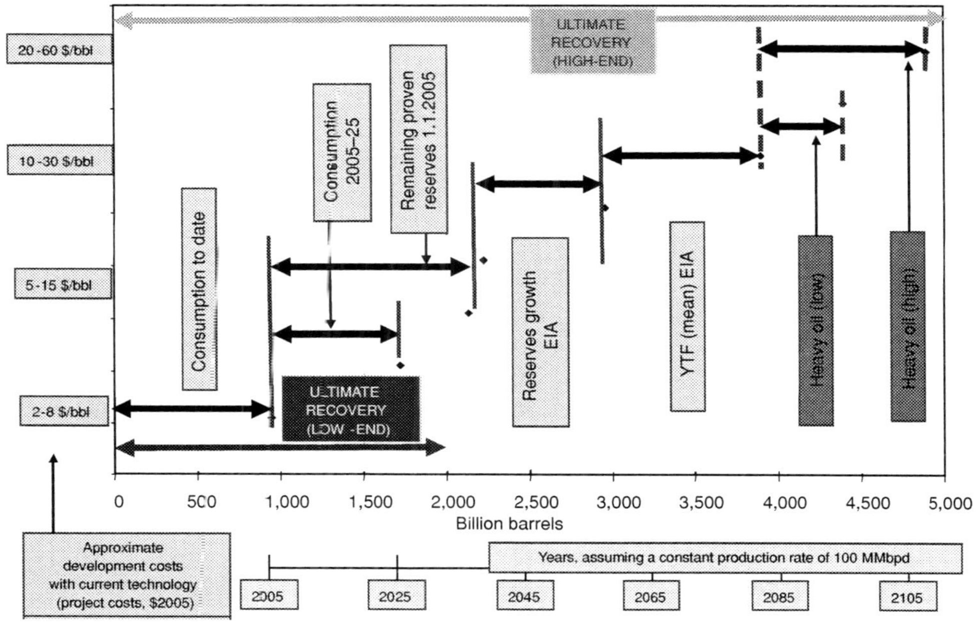

Source: Repsol YPF

Figure 6.13 Summary of Volumetric Estimates

different continents: Asia, Europe and North America. In all, the climate and attitude towards the debate on the future of oil were quite different, ranging from the deep concern of the energy-dependent Europeans to the pragmatic attitude found in South East Asia and the positive "can do" approach of the North Americans. At this stage, my opinion can be summarized as follows:

- There is no clear evidence of a shortage of hydrocarbon resources in the near to medium term. There is, however, evidence for the following:
  - The fading out of cheap oil, particularly outside the Middle East
  - Widespread uncertainty over timing, capital requirements and potential resources in key areas (plus geopolitical uncertainties)
  - Strong increased demand
- These three factors imply a wild "roller coaster" market of higher average prices than the previous decade.
- A reliable and comprehensive inventory of reserves, resources and YTF is in everyone's best interest.

Acknowledgments

I would like to thank Dr. Ralf Boscheck, from the International Institute for Management Development (IMD) for the invitation to participate in the very fruitful Energy Futures conference. I am also grateful to Repsol YPF, who sponsored my participation in this event and granted permission to publish this chapter. Thanks also to my colleague Greg Smith for a careful review of the manuscript.

Notes

1. R.O. Baker., E.S.W. Jong and C. Virués, *Understanding the Enigma of Reserves Growth: The Whys*, The Petroleum Society's 5th Canadian International Petroleum Conference (55th Annual Technical Meeting), Calgary, Alberta, Canada, June 8–10, 2004.
2. Minerals Management Service, *Deepwater Gulf of Mexico 2004: America's Expanding Frontier*, US Department of the Interior, MMS, Gulf of Mexico OCS Region, New Orleans, May 2004.
3. British Petroleum, *Statistical Review of World Energy June 2005*, London, June 2005.
4. Michael C. Lynch, *Scientific (and Non-Scientific) Petroleum Supply Forecasting*, Offshore Technology Conference, Houston, Texas, May 2004.

5. Energy Information Administration, *International Energy Outlook 2005*, Office of Integrated Analysis and Forecasting, EIA, US Department of Energy, Washington, D.C., 2005.
6. J. Laherrere, *How to Estimate Future Oil Supply and Oil Demand?* International Conference on Oil Demand, Production and Costs – Prospects for the Future, The Danish Technology Council and the Danish Society of Engineers, Copenhagen, December 10, 2003.
7. C. Campbell and J. Zagar, *The Iron Grip of Depletion*, Offshore Technology Conference Paper No. 16863, Malkewicz-Hueni Associates and Association for the Study of Peak Oil and Gas, Houston, Texas, May 6, 2004.
8. B.F. Maxted, *Exploration Opportunities, Challenges and Business Delivery*, APPEX-Upstream Perspectives Forum, Houston, Texas, September 14, 2004.
9. O. Appert, Opening Address of the ASPO Seminar 2003, President of IFP. Paris, May 26, 2003.
10. Merrill Lynch Research, *Producing Oil Markets Research*, Merrill Lynch, Pierce, Fenner & Smith Incorporated (MLPF&S), 2004.
11. For recent and quite contrasting views, see B. Lomborg, *The Skeptical Environmentalist (Measuring the Real State of the World)*, Cambridge University Press, 2005; and M.R. Simmons, *Twilight in the Desert (the Coming Saudi Oil Shock and the World Economy)*, Hoboken, New Jersey: John Wiley & Sons, 2005.

7
Assessing "New" Upstream Business Models

Ralf Boscheck

Observers of current upstream oil and gas developments often equate a change in a company's strategy with a fundamental shift in its underlying business model. But they hardly ever reason the need for new business models based on an explanation of existing ones, their contexts and required modification. This chapter attempts to do just that. Following a brief discussion of the notion of a business model in general and in the context of upstream oil and gas activities, the chapter describes types of upstream players – global majors, independents, contractors and technical service suppliers, national oil companies and host country governments – and the characteristic challenges each of them face. It then proceeds, for each group, to discuss the case of a particular strategy that may be considered to be novel and mold-breaking. An admittedly cursory assessment of the strategies pursued by BP, Spinnaker, Anadarko, Baker Energy, Petrofac, Halliburton and Schlumberger, as well as the UK government in managing the UKCS fiscal regime, does not identify any new or need for new business models. In fact, players seem to develop their given models in line with fairly conventional and even predictable patterns of industry evolution.

Introduction

By mid 2005, strong commodity prices, outpacing increases in production costs and taxes, continued to lift producers' net income and dividends and translated into a wholesale repurchasing of shares and retirement of debt. While not a single oil company had made the 2000 line-up of the Fortune 100 Fastest Growing Companies, the 2005 roster featured 19 of them, including exploration and production (E&P) companies such as Apache (no. 99), oil tanker operator Overseas Ship-holding (no. 45) and independent refiner Valero (no. 19). But there were reasons for concern.

Global production growth had exceeded reserves growth for four years in a row and development costs were rising sharply. Major oil companies were announcing their need for more and bigger projects to meet future demand, but exploration spending industry-wide was still at 1998 levels. In 2004, the average ROE and ROCE of the top 25 independents rose to 20.3% from 13.7% and to 15% from 11% respectively, but while acquisition outlays had gone up, exploration spend had remained flat and capital expenses had been cut by $4.6 billion. International exploration appeared increasingly to be borne by smaller independents. Also, between 1995 and 2005, new technologies applied to the reassessment of known fields had added 75% of oil and gas reserves; others had the potential to raise the field recovery level from the current level of 35% to an average of 45%. Yet, super-majors seemed reluctant to foster innovation or to risk a failed field trial, let alone pay the price for a major technology push. Meanwhile technology owners in the contracting and service sectors, seeing their margins squeezed, explored new ways to market. Finally, three additional factors had begun to change the broader competitive landscape: the arrival of national oil companies as powerful competitors for international E&P acreage; changes in access, taxes and operating conditions in a number of major host countries; and the need to rejuvenate maturing regions. In any of these, potential changes in oil prices figured prominently. In September 2005, futures prices had gone past the $60 per barrel mark while some majors continued to assume that $25 to $30 per barrel was the mid-cycle range.

Speculating about how the industry would cope with these and related issues, most observers are quick to call for new business models to be found but often fail to explain existing ones, their contexts and the needs for modification. This chapter attempts to do just that. By way of introduction, the first section briefly discusses the notion of a business model and offers some considerations for its application to the assessment of upstream oil and gas activities. The following section characterizes types of upstream players – global majors, independents, contractors and technical service suppliers – and the major challenges they currently face. The third section discusses some public policy considerations for managing and adjusting a country's hydrocarbon economy and the potential role of national oil companies. The final section presents a summary and conclusions.

Thinking in terms of business models in the context of upstream oil and gas

Shaping one's competitive environment requires a clear understanding of the interaction of various factors driving it as well as of one's own ability to contribute and capture value. Business models distill the essence of this for effective communication, execution and review. Despite being the subject of vast amounts of business and economic research, business models tend to be presented in a rather fragmented fashion, focusing either on the level of activities, business units, corporate portfolios or industrial clusters, and from either a business or public policy angle.[1] In assessing upstream oil and gas markets, a more integrative perspective is clearly required. Research at IMD has resulted in an economics-based framework that links various levels of analysis and challenges the consistency of business models in pursuing private or public objectives. It tests the definition and attractiveness of a given market, operational and investment decisions for reaching it, and issues of coordination, organization and sustainability. Applied to upstream oil and gas, three aspects need to be considered:

First, characterizing a given resource base "play" requires an assessment of its fiscal regime, political context, operating conditions, geology and materiality. Fiscal regimes, involving royalty/tax systems, production sharing agreements (PSAs) or service contracts, differ in terms of title transfer, government take and participation, ringfencing, and lifting entitlement. Hence, they offer different risks and rewards as well as varying levels of expected stability. Political conditions can be gauged by appraising the process of interest articulation and representation and macroeconomic fundamentals. Operating environments differ most importantly according to location (onshore vs offshore, deep vs shallow water etc.), seasonal restrictions, availability and cost of data, infrastructure conditions, drilling costs and outsourcing opportunities. Geology drives the probability of success, the minimum required reserve as well as the required number and locations of wells and supporting infrastructures. Materiality equates to the total estimated resource potential of a given quality. However any given play will only be considered by those able to coordinate operational requirements involved in entry, exploration, appraisal, development, production and ultimately sales and transport. Hence, the attractiveness of any play is a function of the above characteristics, and the number of suitors willing to offer the adequate resource mix, i.e. the substitution potential for demand and supply.

Second, entry implies that operational roles need to be assumed. Figure 7.1 identifies 45 separable tasks, not including finance, throughout the upstream business system that need to be brought together by either internal organization or external coordination.[2] Subject to licensing conditions, outsourcing may be an option if adequate operators, contractors and service partners are available, markets are liquid, switching costs are limited and tasks can be controlled. In any other case, tasks will need to be performed internally and their underlying assets, capabilities and systems embedded in a supporting organization. All of these decisions, including entry, require review to reflect changes in the economics of operation, the maturity of the field, regulatory and technological contexts, commodity prices, conditions in supplier and capital markets, the portfolio of investments and opportunity costs.

Third, there is *a priori* no reason why private or governmental ownership should result in different views on the attractiveness of a play and options for operations. In fact, new public management (NPM) tends to mimic highly contestable market incentives;[3] conversely, a proper fiscal and regulatory regime may substantially reduce or even obviate the need for any direct governmental involvement. Still, national oil companies may be used to pursue non-commercial objectives.

Business models capture the economic logic for aligning internal decisions in view of external conditions. They are explanatory not predictive tools of management. They are unlikely to reflect the specifics of any given circumstance, but rather serve as blueprints for testing the consistency, sustainability and required evolution of a given case. The following two sections do this for different types of players and plays.

Players: majors, independents and contractor/technical service firms

Players in upstream oil and gas are typically grouped as integrated, independent or national oil companies as well as contract/service firms. Of the first, five global "majors" (Exxon-Mobil, BP, Shell, Chevron-Texaco and Total) accounted for 17% of the world market in 2004 and cover the full cycle from upstream exploration through transport and storage to marketing and sales of refined products – from plastics to gasoline. The typical independent operates exclusively in the upstream exploration and production segment, with no refining or marketing operations, and ranges in size from one-or-two person companies with small turnovers to large publicly traded "super-independents" with billion dollar sales. A subgroup of independents, the so-called exploration boutiques, focuses

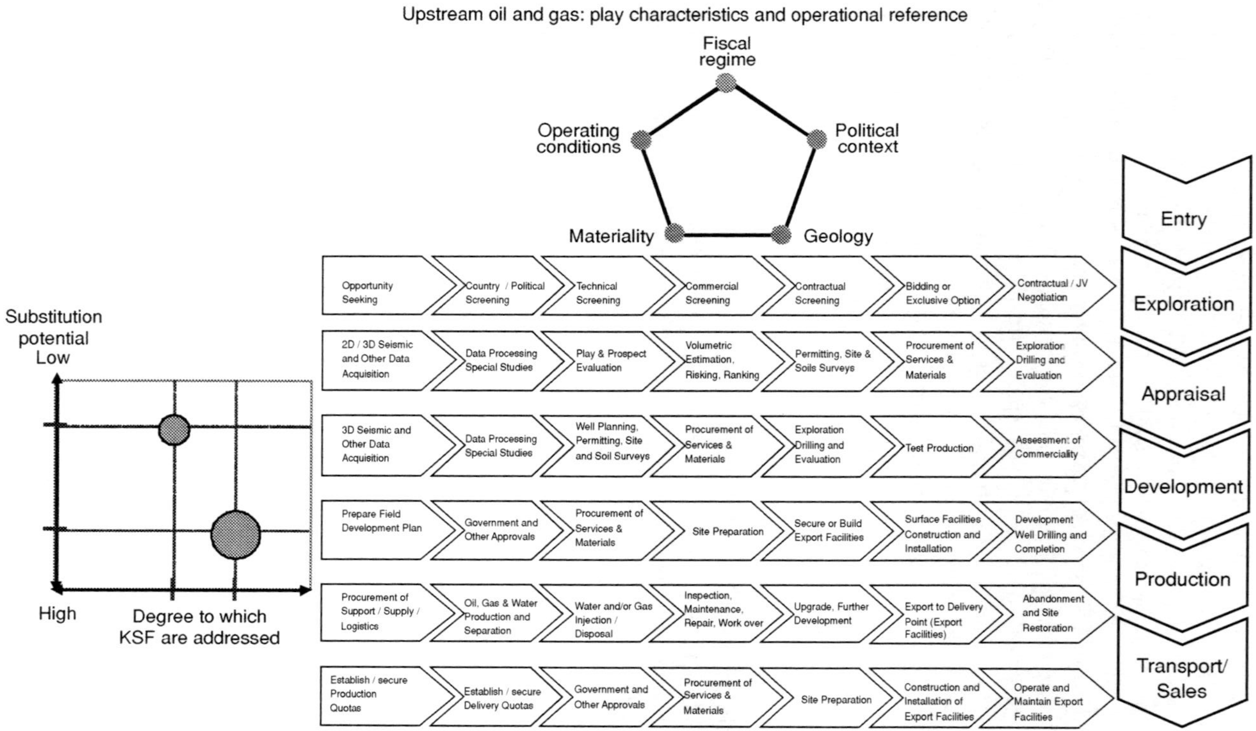

Figure 7.1 Elements of an Upstream Business Model

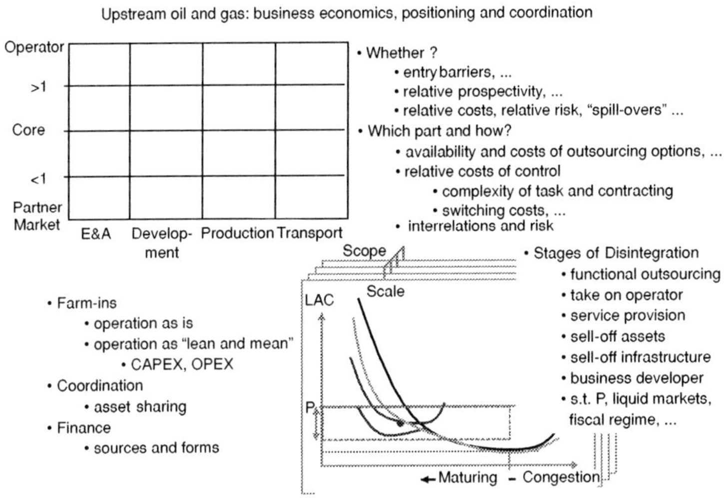

Figure 7.2 Elements of an Upstream Business Model

entirely on exploration based on a small number of licenses, raises venture capital to begin small-scale operations and tends to be either acquired, if successful, or liquidated, if not. National oil companies accounted for 30% of the world market in 2004 and have persistently grown their technical and commercial capabilities; with strong home government support many are expanding beyond their borders in order to compete. Technology firms and engineering contractors respond to oil companies' varied outsourcing needs. Subject to client demands and market conditions, their offers cover virtually all areas of exploration and production (E&P) and a variety of performance and financial contracts. Focusing on global majors, independents, technical service firms and contractors, the following subsections outline their apparent business models and then illustrates each with an actual example.

Global majors

Characteristics and challenges

Since the very beginning of the modern petroleum industry in the late 1860s in the United States, large-scale, high-risk investments, non-existing markets for logistics and technological services and unsteady supplier relations had driven operators to seek control over all upstream and downstream activities. Vertical integration soon triggered horizontal consolidation until in 1911 the US Supreme Court dissolved Standard Oil Trusts into 34 separate entities. In the late 1940s, two of the descendent

Table 7.1 Key Statistics by Type of Industry Player

	Global majors					Independents		
	BP	Exxon-Mobil	Royal Dutch Shell	Chevron-Texaco	Total	Spinnaker	Anadarko	Toreador
Upstream Revenues	$19.8bn	$16.7bn	$39.6bn	$9.5bn	$36.2bn	$272.9m	$6,067.0m	$22.4m
Oil: b/d	2.7m	2.9m	2.5m	2.4m	1.7m	1,743	310,000	1,735
Gas: boe/d	1.6m	2.0m	1.7m	0.8m	0.9m	5,955	370,000	n.a.
Market cap.	$227.8bn	$349.1bn	$210.1bn	$117.3bn	$150.6bn	$2.2bn	$20.4bn	$0.4bn
No. of employees	102,900	85,900	112,000	56,000	111,401	78	3,300	45
Upstream revenue/employee	$192,420	$194,412	$353,571	$169,643	$324,979	$3,498,564	$1,838,485	$496,667
Replacement cost: $ per boe	4.44	5.56	11.97	6.43	5.24	n.a.	8.93	4.54

Note: For Exxon, upstream revenues include exploration, development, production, and gas and power marketing. Total's reported E&P revenues include inter-group sales.

Source: Petroleum Economist, Annual Reports 2004, Thomson Financial

firms, Exxon, formerly Standard Oil of New Jersey, and Amoco, formerly Standard Oil of Indiana, joined Mobil Oil, Royal Dutch Shell, BP (British Anglo Persian Oil Company), Texaco (Texas Fuel Company), and Chevron (including Gulf Oil) in a group known as the Seven Sisters. Operating as interlocking ventures to build negotiation power and share risk and finance, these companies grew fast with the rise of international oil production until, after the wave of nationalizations in the Middle East and South America, every single one of them had to be rebuilt. A series of mergers and acquisitions helped the major companies regain their global scale and scope in operations from exploration to retailing: Exxon teamed up with Mobil, Chevron with Texaco, and BP with Amoco and Arco. In France, Total joined arch-rival Elf Aquitaine. But only by 2004 had BP's production of 4 million barrels per day reached its 1975 level.

For more than 140 years, vertical integration in the petroleum industry had offered the means to balance capacity utilization, ensure market access and, together with geographic and product diversification and public share ownership, spread risk. Born out of necessity, systems growth fed increased operational scale, drove up the need to replace reserves at ever-larger rates and the funding needed to fuel exploration. Are current operating and market conditions sufficiently distinct to seriously challenge the majors' approach and require a new business model? The primary concerns relate to market access, risk aversion, coordination, and bureaucratization.

First, although global majors operate on every continent, mature regions of North America and northwestern Europe account for about 60% of their current oil and natural gas production. In those areas, declining output raises production costs[4] and makes justifying additional investments difficult. Opportunities for growth come at a price. Participation in new product markets – from gas-to-liquids (GTL) and liquefied natural gas (LNG) to oil-sands and heavy oils – calls for large-scale investments in processing and transport. In addition, many of the world's remaining conventional sources of oil and natural gas are located in technically challenging regions, such as the Arctic and Asia-Pacific, or in countries with relatively high political and legal instability, such as Nigeria, Russia or Sudan. Since the late 1990s, when several large finds in deepwater locations raised global reserves, the average size of new discoveries around the world has declined threefold, to about 22 million barrels of oil. Also, majors are often unwilling or unable to operate in highly sensitive regions.[5] How does a typical major respond to these growth challenges?

Table 7.2 Key Statistics, Top 3 Majors

Global majors (as of 2005)	Share of liquid production '000 b/d	Share of gas production '000 boe/d	Share of commercial reserves (total oil eq.) mmboe	Production per employee boe/employee	Market Cap. US$ mio	Sales US$ mio	No. of employees	Credit Rating S&P
Total World	73,967	46,198	1,277,702	n/a	n/a	n/a	n/a	n/a
BP plc	2,735	1,580	33,084	27	227,753	297,507	102,900	AA+
Exxon-Mobil	2,906	1,960	40,259	34	349,115	263,989	85,900	AAA
Shell	2,541	1,686	29,485	23	210,080	276,770	112,000	AA

Notes: (1) Liquids = oil, LPG and condensate.
(2) Commercial reserves = reserves in production, under development or likely to be approved for development within 2–3 years. Roughly equivalent to company Proved + Probable reserves.

Source: Petroleum Economist, Thomson Financial

Second, mergers helped majors to replace their reserves at below drill-bit costs and to expand exploration portfolios with lower-risk prospects. Yet, increasing the scale of operations makes it ever more difficult to find fields large enough to make a real difference. No wonder therefore that in a sustained high oil price environment, players who would normally seek to rationalize the tail-end of their portfolios at the top of the market continue to operate non-core, marginal areas either directly or under some form of outsourcing arrangement.[6] Does this behavior signify risk aversion? Is it necessary for majors to get around avoiding risks? If so, how can management attention be refocused on capturing true growth opportunities?

Third, future growth requires a combination of acquisitions, exploration activities, and further technology and downstream market developments. Global majors, seen by their investors to hold excellent medium-term growth potential, currently may have little problem financing these. Yet spending these funds in the most efficient way requires companies to prioritize the establishment and control of future, large-scale essential infrastructures – from pipelines to re-gasification and liquefaction plants to hydro-crackers – while leveraging available outside parties for operational tasks wherever possible. In the process, controlling cooperation remains vital. Hence, with high oil prices putting a premium on acquisitions and increasing the value of an exploration upside, majors need to think about "recreating" their dedicated venture teams to find and secure new exploration opportunities *and* complement their work by taking a stake in, or even partner with, small but tightly focused exploration companies. Similarly, in the area of technology, majors need to step up their in-house investments while, at the same time, cooperating with technology hopefuls and vendors.[7] But would opening up to more extensive market-based coordination and networking fundamentally change the coordination logic underlying the majors' business model?

Fourth, operating on a global scale and under significant technological, political and economic risk, as well as public scrutiny, calls for comprehensive systems of external and internal supervision. All global majors need to operate sophisticated matrix organizations of significant size and make use of public relations departments to ensure that the corporate image is not tarnished by accidental oil spills or political controversy. Yet, there is growing concern that too much focus on governance concerns may drive up bureaucracy at the expense of market responsiveness. Is it possible to maintain major operations, address stakeholder demands and avoid red tape? How does a global major like BP respond to any of the above concerns?

Table 7.3 Key Statistics, BP

	Acquisition costs (proved)	Acquisition costs (unproved)	Exploration costs	Development costs	Total cost incurred	Production	Year end reserves	3-year proved acq. costs $/boe rank	3-year proved dev. costs $/boe rank	3-year replacement costs $/boe rank	Net income $/boe
BP plc	–	$78.00m	$1,838m	$8,706m	$9,822m	$–1,451.5m	$18,471m	2.59	5.28	4.44	8.22
BP plc ranking		69	7	3	3	3	3	27	23	17	117

Source: Adapted from J.S. Herold

BP – Scale, scope and the markets

In 2005 BP looked back on healthy 2004 results, helped by a surge in production growth (nearly 11% upon the previous year), mostly resulting from TNK-BP operations in Russia. Its 2005 pro forma results equaled US$25.7 billion, of which 77% were derived from upstream and 22% from downstream, with some contribution from other segments. Upstream capital expenditure accounted for 65% of capital expenditure in 2004, rising to an expected 68% of US$14.1 billion[8] by year end. Downstream represented around 20% of spending, chemicals around 6% and gas, power and renewables around 4%. The company's current performance was the outcome of a set of deliberate decisions.

When in 2001 BP set out its growth strategy, CEO Lord Brown[9] formulated six objectives: (1) grow in size to undertake large-scale projects that offer a truly distinctive return to finance more development projects, hold costs down and potentially win more reserve auctions; (2) diversify risk; (3) leverage vertical integration but enforce market discipline across the organization; (4) understand change in all parts of the markets and have the skill to manage that change; (5) have a first-mover advantage in picking your partners; and (6) offer a single window to host governments. The resulting actions quite clearly tackle some of the above concerns.

In addressing *growth challenges and maturing plays*, TNK-BP is expected to drive the company's future performance. In addition, upcoming projects in Vietnam, Australia, Indonesia, Azerbaijan, Algeria, Angola, deepwater Gulf of Mexico, Trinidad and Tobago, and Alaska are to help counterbalance losses from mature fields in the North Sea, Egypt and Gulf of Mexico. For the 2004–06 period, the company aims to (1) manage maturing assets and divest itself of non-performing ones; (2) invest in the "largest lowest-cost new hydrocarbon deposits" to keep production growth at 5% from 2004 to 2010; (3) increase LNG and gas production as a share of total hydrocarbon production to around 20% by 2008.[10] BP does not intend to pursue unconventional oil and gas.

In October 2005, the company signed a joint venture with India's Hindustan Petroleum with an initial plan to build a $3 billion refinery, and attempted to buy a substantial equity stake in Sinopec, China's largest refiner and marketer. The deals would give the company a foothold in the markets of the fastest-growing users of energy among large economies, which so far had rejected several such attempts by BP, Exxon and Shell. Still many drawbacks exist, at least in the short term. In both China and India the state exercises strong control over the ownership and pricing of oil and gas products. Yet, for BP this is an acceptable cost for attaining resource and market access and just one more aspect to be considered in pursuing the company's strategy of *partnering and leveraging integration.*

With regard to the former, BP continues to stress the philosophy of partnering to "seek mutual advantage." Two years after the TNK-BP merger, President Putin publicly announced that he was pleased to hear the company being called "a corporate citizen of Russia" and that it took part in social welfare programs. On the same occasion, Lord Brown ensured that the venture was carrying out all four "missions" that Mr. Putin had assigned it before the merger in 2003: increasing oil extraction volumes, introducing new technologies, developing corporate management, and maintaining decency in the company.

As for integration, BP officially downplays the value of vertical integration but its top management nevertheless appreciates "some" vertical control: According to Lord Brown having "(t)he right refineries in the right geographic areas may provide a benefit. Also, shareholders invest in the agency of the management to more easily participate in portions of the industry than they can do themselves. It is easier to invest in BP than obtain a private equity stake in China." And in the words of Chief Financial Officer John Buchanan "(s)ome advantages of vertical integration come from unpredictable regulated markets. You never know where the rent will show up. Markets aren't yet perfect, and there is not contestability at each point in the value chain."[11]

At the same time, there needs to be *a check on potential inefficiencies due to integration.* In BP, the absence of purchasing commitments and transfer pricing rules that mimic market mechanisms are to "expose intra-business subsidies and force the businesses to become more competitive." Also, the market figures strongly in offsetting another potential drawback of in-house scale – *red tape*. BP business unit leaders' careers and compensation are driven by annual "performance contracts" specifying cash flow, net income, and new investment targets; they periodically meet in peer groups to benchmark each other against a list of operational key performance indicators. In addition, the top quartile performers are charged with improving the bottom quartile's results.[12] Furthermore, performance targets, such as the company's highly publicized unilateral commitment to cut its emissions of greenhouse gases by 10% from a 1990 baseline, are set to fuel internal competition and have been shown to markedly improve innovation and productivity.

Hence, while relative to the early years of the oil industry, current market and outsourcing opportunities may make continued global integration appear anachronistic, BP is set not only on maintaining its size but on increasing it even further. The company uses its balance sheet to finance otherwise unobtainable large-scale infrastructures, and leverages its political and economic network to pre-empt markets and resource access. Its diversity smoothes income over the economic cycle. It is an extension of "big oil" at a time when smaller, non-integrated independents appear to be pursuing a very different business model.

Independents

Characteristics and challenges[13]

Originating in the North American and North Sea markets, a by now global group of non-integrated exploration and production companies is starkly polarized between a few "super-independents" and myriad small ones. While companies such as Anadarko, Apache, or Talisman Energy seem set to join "the big league" in terms of geographic scope, technological complexity and spend, the typical US independent is still a largely US-focused small business owner who has been established for many years.[14] International projects, if any, are mainly onshore E&P and production enhancement, with the biggest hindrances to international operations being capital expenses and political uncertainty. The fact that Northern European independents are on average bigger and more international than their US counterparts reflects differences in both the size of the underlying population and the history of the sectors in each region.[15] Still, in addition to differences in origin and size, the diverse nature of an independent's operations – from exploration, to development, to production – at different stages of field and industry development – from prospecting to large scale exploitation within an existing infrastructure to maintaining a low capacity stripper well – makes typecasting independents not only conceptually difficult but empirically hard to substantiate. And yet there are some intuitively appealing, size and investment related characteristics that industry members, financial analysts and researchers typically associate with an independent E&P company. These elements of an independent's business model – related to heritage, funding, market-choice, decision-making and risk-taking – may come under pressure as the company grows and its environment changes.

First, three forms of operations are often seen as the start of an independent: exploration leg-work, farm-ins or consolidation. With regard to the first, smaller companies typically undertake geological studies, seismic testing and early government negotiations and then put together consortia of larger companies to do the actual exploration. In return, the smaller partner receives a carried interest through the first two or three exploration wells, and in case of discovery, expects to participate in the development or production phase covered by the group's anticipated project financing. In the second case, an independent player takes over acreage from larger oil companies in fields that are past their prime or in small, marginal fields that are deemed less viable by the big operators. Finally, consolidation involves financing, acquiring and combining operations to share costs and restructure asset portfolios.

Second, funding any of these ventures calls for salesmanship with banks, partners and investors. Start-ups with no track record typically require private capital funds, vendor participation or a joint venture with industry partners. For others, internally generated capital may be complemented by private-equity funding and leveraged by cheaper forms of non-recourse debt financing (possibly involving negative pledges). In either case, taking over operatorship within a partnership improves cost control and reduces financing costs. Volumetric production payments (VPPs) – i.e. presales of a fixed proportion of future production at an agreed price – offer cash up front. Mezzanine financing, combining elements of a junk loan with a retained equity portion, is more expensive as it compensates the lender for extra risk. In any event, investors and borrowers require collateral. Proved reserves are the only reserve classification that the US Securities and Exchange Commission (SEC) allows a publicly traded company to consider as assets, but their value may fluctuate depending on operating costs and commodity prices.[16] The sale of equity is often the ultimate goal of an oil and gas company, and other forms of capital are used as a means to achieve this goal.

Third, in choosing their markets and growing their operations, British and American independents have been found to follow five types of approaches:[17] (a) country focused operations seek to leverage local knowledge and goodwill; (b) mature asset specialists operate many geographically concentrated licenses and concentrate on invigorating declining plays; (c) regionally focused operators exhibit moderate levels of international license dispersion; (d) internationally diversified independents hold the most geographically dispersed operating licensing; and (e) portfolios have more than 90% of their licenses as non-operators. More important, and statistically significant, however, are two findings related to relative size, growth and effectiveness.

Fourth, smaller independents appear to have an advantage over majors in decision-making speed, the authority to commit, the seniority of personnel and the establishment of relationships. They are thus more successful in applying for licenses in non-Western countries and are able to target non-traditional, under-explored overseas areas. In fact, some display an active willingness to engage with prospective countries, where due to relatively high political and/or security risks competition

Table 7.4 Key Statistics, Independents

Independents (as of 2005)	Share of liquid production '000 b/d	Share of gas production '000 boe/d	Share of commercial reserves (total oil eq.) mmboe	Production per employee boe/employee	Sales US$ mio	Net Income US$ mio	Market Cap. US$ mio	No. of employees	Credit Rating S&P
Total World	73,967	46,198	1,277,702	n/a	n/a	n/a	n/a	n/a	n/a
Andarko Petroleum Corp.	310	370	3,694	94	6,067	1,606	20,382	3,300	BBB+
Apache Corp.	310	285	3,792	117	5,333	1,669	19,741	2.642	A–
Encaca Corp.	313	585	6,390	77	12,723	3,785	41,137	4,090	A–
Pogo Producing Co.*	51	54	116	198	1,312	262	3,010	260	B+
Progress Energy Trust	3	10	54	n/a	134	44	883	n/a	n/a
Spinnaker Exploration Co.	2	6	51	22	273	54	2,204	78	n/a
Toreador Resources Corp.	2	n/a	14	39	22	25	385	45	n/a

Notes:: (1) Liquids = oil, LPG and condensate.
(2) Commercial reserves = reserves in production, under development or likely to be approved for development within 2–3 years. Roughly equivalent to company Proved + Probable reserves.
* Pogo Producing senior unsecured debt not rated. Rating given is for a September 2005 senior subordinated debt issue.

Source: Petroleum Economist, Thomson Financial, company report

for licenses is reduced. Yet, while speed and the absence of a consumer/investor brand allow small independents to exploit the low risk tolerance of integrated players, super-independents are increasingly facing growth problems similar to those of global majors. Looking at Devon, Apache, Anadarko, Burlington Resources, Kerr-McGee, Amerada Hess and Unocal, Farley[18] found in 2003 that the group averaged 15% annualized returns when growing their enterprise value from $5 billion to $10 billion. But the group's average annualized returns fell to a negative 5% when the value exceeded $10 billion. Do companies face an apparent $10 billion enterprise value threshold because it is difficult for them to find, acquire, and exploit opportunities of substantive scale, or because they fail to maintain their entrepreneurial drive and cost control? How will independents of different sizes and origins deal with changes in oil prices that either shrink margins or conversely keep majors holding on to marginal assets? How will they deal with increased costs of capital?

Independent oil companies, as a group, suffered when in 1985 low oil prices exposed their lack of financial strength and management talent. Many simply disappeared. They experienced a renaissance throughout the 1990s when new technologies, such as 3D seismic and horizontal drilling, helped lower finding or development costs, sales of marginal fields allowed farm-ins, and funding and outsourcing opportunities were readily available. Nowadays, funding seems not to be an issue, and the typical independent's debt capacity has greatly improved. Yet, while some independents deploy large amounts of E&P capital into more drilling, others turn to lower risk methods of growth that offer limited production but far less strain on the balance sheet. Consider the cases of Spinnaker Exploration and Anadarko Petroleum.

Spinnaker Exploration – financial venturing Founded in 1997, Spinnaker, an independent energy company recognized for its strength in data integration and seismic image enhancement, is engaged in exploration, development and production of natural gas and oil mostly in the Gulf of Mexico (GoM). Beginning as a conventional shelf player the company was soon drawn into deepwater exploration and, by 2004, held approximately 131 blocks covering 745,000 gross acres, or 312,000 net acres, in the deep water of the GoM. At the same time, it had discovered 11 new deepwater fields, two of which are currently producing, four are in development and five more still under evaluation. Spinnaker aimed to offer investors a balanced inventory of conventional and deep shelf activities in various regions of the GoM and, in 2004, increased geographic diversification by venturing internationally to Nigeria's Tari prospect. It is the financial aspect of Spinnaker's growth that is of importance here.

At the time of formation, Spinnaker was the largest equity investment into any grass-roots energy start-up – ever. Petroleum Geo-Services ASA (PGS) had put up $15 million capital and granted Spinnaker a perpetual license to use PGS's existing non-proprietary 3D marine seismic data and any additional non-proprietary seismic data that PGS produced until April 2002. To balance out PGS's commitment, Spinnaker received $60 million private equity from E.M. Warburg, Pincus & Co. LLC, New York that saw an exploration and production potential big enough to access the public markets. The next stages of financing involved an $85 million conventional line of credit based upon the strong balance sheets of Spinnaker's partners, an initial public offering in September 1999, and a secondary offering of $140 million in net proceeds in August 2000. In late 2000, PGS sold all its holdings in the company via a registered offering for $150 million in net proceeds. As of September 30, 2001, Warburg owned 26% interest in Spinnaker, management owned 13%, and the public owned 61%. Spinnaker had no debt. A foundation had been laid to fuel the company's exploration efforts through successive public offerings of preferred and common stocks and debt securities etc. as well as conventional lines of credits with banking consortia.

In September 2005, Norsk Hydro acquired Spinnaker and its fairly modest 62 million boe of proven reserves for $2.56 billion, i.e. for $41.29 per boe.[19] Following rival Statoil, which in April 2005 had already replaced reserves and a declining home production with GoM deepwater assets, Norsk Hydro radically shifted price assumptions, applying a long-term oil price of around $35 per barrel to value Spinnaker. By comparison, Chevron had acquired Unocal for $8.62/boe of proven reserves and Statoil's acquisition of EnCana's US GoM portfolio was valued at $5.99/boe of proven and probable reserves. One may question the willingness of other buyers to follow Norsk Hydro's price assumptions; in which case, Spinnaker's timing was right and its exit was very much in line with its existing business model.

Anadarko Petroleum – jumping the growth hurdle? Set up in 1959 by the Panhandle Eastern Pipeline Company to find and extract natural gas in the Anadarko Basin of Texas, Oklahoma and Kansas, Anadarko Petroleum expanded in the Gulf of Mexico before being spun off in 1986. Over the next decade the company became one of the largest independent E&P firms in the USA. In the late 1990s, low oil and gas prices and rising finding and development costs suppressed financial returns and wanting organic reserve additions obliged Anadarko to make expensive acquisitions in order to grow. A larger reserve base however made aggressive growth expectations harder to achieve and increasingly pushed Anadarko to compete with the majors for opportunities big enough to enable the company to grow. Rising oil prices at the beginning of the new century did not solve Anadarko's problems – they just made them more visible. The company found itself needing to restructure its assets and recreate an entrepreneurial culture.[20]

By 2003, a new management team was reviewing the company's properties based on operating costs, remaining growth opportunities, margins, decline rates, rates of return, free cash flow, manpower, skill requirements, and general and administration (G&A) expenses. It was found that around 90% of Anadarko's value was generated by 25% of its fields. Hence the company identified scattered, ungrowable or declining properties for divestment, and retained sustainable, low maintenance output generators to fuel growth. Management decided to concentrate on its North American onshore asset base to maximize cash and invest it in international and domestic frontier areas. Since in the process, getting rid of ungrowable or scattered yet very profitable wells would dilute earnings per share, stock repurchasing and the retirement of debt were expected to keep share prices up and strengthen the balance sheet. In June 2004, CEO Jim Hackett announced plans to sell 11% of Anadarko's proven reserves and about 25% of production – about 70% of its fields – for anticipated after tax proceeds of over $2.5 billion. The $2.8 billion after tax actually generated was used first to reduce net debt by $1.5 billion and then to return value to shareholders via a share repurchase of $1.3 billion.

Clearly, one may wonder whether shrinking to grow is the proper strategy to follow in a high oil price environment in which delaying asset sales increases value, repurchasing shares becomes ever more expensive, and an increasing stock value permits more debt to be taken. One may also question Hackett's decision given Anadarko's ambition to take on major oil companies by becoming more global, and expanding into more deepwater drilling and LNG. Hackett would respond "(w)e want to be able to withstand downturns in oil and gas prices without giving up the ability to take on new opportunities when they become available. The best time to fix a roof is when the sun is shining."[21] For Hackett this was reason enough to tackle internal structures too.

Over the years, Anadarko's top management style had not kept up with the company's growth. The original entrepreneurial approach had succumbed to top-down communication with little consultation and engagement from below, had spread hierarchical thinking, proliferated teams and ultimately dampened any sense of accountability. In the words of an observer "(p)eople have been here for a long time, which created a mentality like working for the government that manifested itself in the administrative side of the business as a culture that existed to service whatever the E&P side wanted. That meant not pushing back when it came to cost."[22] Hackett adopted four measures to change this: (1) APEX – achieving performance excellence – introducing performance contracts, 360-degree feedbacks for every employee and regular meetings for management communication and buy in; (2) streamlining of management – increasing the number of reports per manager, collapsing parallel groups into one, reducing the number of technical and executive posts; (3) announcing the first lay-off in the history of the company; and (4) asking executives to follow his model and have lunch with everyone else in the staff cafeteria.

Despite the difficulties in typecasting independents mentioned above, it is possible to point to some intuitively obvious patterns of funding, market-choice, decision-making and risk-taking that are likely to change as companies grow and their environment changes. Spinnaker and Anadarko follow the same business model of a non-integrated E&P company. But both are at different stages of a company's evolution and therefore face different challenges and options for responding to those challenges. Spinnaker focused on attracting finance to offer a portfolio of conventional and increasingly deep-shelf activities for acquisition by a bigger player ready to leverage larger internal finance and management scale. To quote Richard Anderson, CEO of First Calgary, another young independent: "Most companies are built to be sold and we're probably no exception." Conversely, Anadarko's decision to shrink for growth and shake up the internal organization reflects prudent housekeeping of a major player in a cyclical business and no downstream diversification opportunities. Super-independents require cost-control and maximum use of outsourcing opportunities; whether they are able to crack the $10 billion mark and enter the realm of global majors is unclear – as is the motivation for trying to do so.

Technical services and contractors

Characteristics and challenges

By 2005, sustained oil price increases finally lifted the profitability and capacity utilization of contractors and service companies. And yet in 2003/04, around five years after oil prices had rebounded in 1999, contractors and service companies remained under pressure from super-majors determined to control costs and maximize their profits. In 2002, when BP increased its return on capital employed (ROCE) by 23%, Elf Exploration UK by 27.7%, Shell by 17.2% and Total E&P by 36.1%, the ROCE of Schlumberger, Halliburton, Smith International and Baker Hughes had slumped by 22%, 16%, 6% and 2% respectively. For the same period in Europe, Stolt's ROCE figure was down by 14%. Contractors and technical service firms reacted by becoming more client-responsive and in the process combined their offers.

Oil companies, downsizing their technical staff, were increasingly looking for outsourcing opportunities and demanded single contacts and integrated services – from technical services to operations and maintenance to asset sharing programs. Smaller suppliers had to sell out and the industry consolidated. Halliburton merged with Dresser, Schlumberger with Camco, and Baker Hughes with Western Atlas. Consolidation also affected engineering contractors and drilling companies. Where previously the oil majors had undertaken a large share of the industry's R&D, suppliers now took over. During the late 1990s, the decision by most of the super-majors to react to eroding oil prices by cutting back their technology spend had an immediate impact on US intellectual property filings. In 1997, the big-six (US) service companies (Weatherford, Baker, Schlumberger, Halliburton, BJ Services and Smith International) were granted about 135 upstream patents, compared with some 35 for Shell, BP, Exxon-Mobil and Chevron-Texaco. By 2002, that gap had widened to 400 for the service sector as opposed to only 30 for the super-majors. Clearly, there was great potential for contractor and technical service companies to develop technology, expand their offerings and bring them to market. In the process, some changed their business and others their way of contracting. Consider the following cases:

Tetra Technologies/Maritech – from contractor to independent The US GoM currently houses an estimated 4,200 structures with an average age of over 20 years, near or past their intended design life. According to the Mineral Management Services (MMS) operators will spend $10 billion over the next decade on well abandonment and decommissioning (WA&D) services. The MMS obliges operators to plug and abandon wells and remove platforms and flowlines within one year of a lease expiring; in addition, the Financial Accounting Standards Board (FASB 143) requires companies to recognize abandonment obligations as a separate liability on the balance sheet. Hence, operators are looking for an opportunity to sell mature fields to an undertaking that specializes and has a track record in abandonment and decommissioning, as well as a strong financial base to limit the bankruptcy risk.

Seeing an opportunity, Tetra Technologies Inc.,[23] a provider of traditional WA&D services, set up Maritech Resources Inc. with two goals in mind: 1) To buy properties from operators, who thereafter would be completely freed of regulatory liabilities, somewhat earlier in their lives when there is enough remaining reserve value to pay for some or all of the abandonment costs. 2) To accumulate a large backlog of abandonment work for Tetra, which takes over after remaining field reserves have been completed, to improve Tetra capacity utilization and work schedule. Success demanded skills in reserve valuation, superior skills in resource exploitation, stability of cost and regulatory requirements, and ability to hedge oil price changes. Maritech started in 2000 with just seven operated fields; by August 2005, it operated 34. In the past five years, the company has plugged over 120 wells and removed 40 structures. A solid track record – poised to grow. Is this a new business model? No, it is an alignment of two businesses through joint ownership. Other players seek alignment through contracts and incentives.

Baker Energy, Petrofac, Halliburton, Schlumberger – aligning incentives
Contractors and technical service continuously expand their offers and ways to incentivize performance. In doing so they reduce barriers to entry for new oil competitors and at times can be seen to be infringing on the business of their current customers. Far from being new, some of these approaches are mere aberrations of the existing business model, carrying substantial financial and strategic risks, and are therefore only sparingly used.

Baker Energy, a provider of operations and maintenance services, acts as the network manager, consolidating production operations between the wellhead and the sales meter for multiple companies. Participating companies share resources such as helicopters, marine vessels, computerized management systems and technical specialists, and monitor Baker Energy's performance according to previously agreed measures. Baker Energy determines how to deliver the results, making daily operational decisions such as supply vessel routing and capacity utilization or maintenance planning. For assuming this additional responsibility, Baker Energy is eligible to receive a share of the cost savings it achieves, as well as a performance bonus. Total production operated by Baker Energy's so-called OPCO network is currently in the top ten of all Gulf of Mexico producers.

Petrofac offers engineering and construction as well as operational and facility management services in the North Sea. It is currently being used in a range of assignments as a central duty holder, leaving the responsibility of operatorship with the client – typically an independent or major oil company. The company aims to develop as turnkey operator, allowing oil companies to focus on reservoir and investment management. Where Petrofac's engineering, construction or operational services are likely to drive major value for the client, the company seeks a minority co-investment to align interest.

Halliburton and *Schlumberger*, world leading providers of wireline, drilling, well and lifting services, are extending their offer into what Schlumberger calls integrated project management (IPM). The IPM business offers a complete suite of consulting, engineering and execution services and is typically priced on a fixed-price "turnkey" basis, with most project execution risk accepted by the provider. The program addresses the needs of national oil companies with only modest geo-technical and engineering manpower and/or producers new to a region. Analysts maintain that "given high factor utilization and rising prices Schlumberger is pursuing less IPM work as its people and equipment are fully occupied."[24] It could also be that IPM-type arrangements simply stretch the risk limits that a service provider is willing to handle. Contracts for enhancing brownfield operations are typically based on incremental production improvements with service firms not acting as operators. Costs are repaid out of cost oil and any potential saving is shared according to a predefined sliding scale. On the positive side, oil price fluctuations may offer windfall gains, despite overall profit caps. On the negative side, in dealing with national oil companies, the service supplier is competing for tasks that were traditionally performed by its major or independent oil company client.

In sum, Figure 7.1 identifies 45 separable tasks throughout the upstream business system that need to be brought together by either internal organization or external coordination. The section on pp. 138–9 outlined the logic for determining a player's scope of engagement and contribution towards a given business model. Global majors and independents differ in terms of their degree of integration and hence diversification opportunities and risk. With the exception of a major's need to protect its downstream brand, their business challenges are quite similar and focus on balancing exploration and acquisition in replacing and growing reserves, increasing operating efficiency, and accessing low-cost finance and prospective resource plays. All independents pursue a non-integrated business model, but decisions and strategies differ most importantly depending on size, age and play. Recent expansions in outsourcing offers presented by technical service firms and contractors largely follow their tried and proven business model. In some cases, however, their usual clients may consider service providers knowingly or unwittingly assuming operatorship a competitive threat. The latter point frequently surfaces in discussions about the changing roles of national oil companies and the part they may play in shaping a country's hydrocarbon economy.

Plays: fiscal regimes and national oil companies (NOCs)

In much the same way as corporate players evaluate the extent of their engagement across the E&P business system, governments face a continuum of options ranging from entirely outsourcing the exploitation of their country's oil and gas resources to completely "internalizing" operations in the hands of a state-owned, state-run enterprise. If there is *a priori* no reason why private or governmental asset ownership would result in different views on the attractiveness of a play and options for operations, why would any country set up its own import-competing NOC rather than creating a regulatory and fiscal regime able to attract global specialists? The answer must lie in the perceived inability of a fiscal system to limit a country's outsourcing risk. Countries may not only perceive this risk differently but that perception may change over time and trigger reactions that, in turn, put companies' operations and business at risk. A number of questions arise from this, which this section will briefly address: What are the main objectives held by host governments and international oil companies in exploiting a resource basin? Is fiscal stability the way to optimize resource utilization? Is there a common thread in the evolution of fiscal regimes around the world? And, finally, what is the role for NOCs in shaping a country's national hydrocarbon business model?

Fiscal regimes and regulatory competition

In determining the structure of its fiscal regime, a host country government typically wants to attract investments and a guaranteed fair share of profits, an assured periodic income, energy provision, technology transfer, employment creation, and efficient resource use. International oil companies, on the other hand, seek profitable prospectivity, resource and market access, transparency, low transaction costs, maximized shareholder return and a diversified portfolio. Both parties face one another in a market; but while objectives are clear outcomes are not.

Ex ante, plays may have to compete for investments by offering the most favorable terms, or, given superior levels of prospectivity, may not need to do this to induce a feeding frenzy among investors. *Ex post*, committed investors are concerned about the stability of the fiscal regime. Clearly, fiscal stability encourages investments, entry and asset trading, sustains risk-taking in exploration, development and production and so ultimately promotes the competitiveness of a basin. But requiring strictly stable fiscal regimes is tantamount to asking for a long-term contract without an adjustment clause. Over time, changes in global commodity prices, investment activities, or the relative prospectivity of the region are likely to change as much as the country's economic policy and view on foreign participation. Systems are necessarily in flux, and, as is shown next, it is the process of adjustment that enhances the competitiveness of a basin.

The evolution of the UK fiscal regime (1963–2005)

Between 2003 and 2004, the UK Continental Shelf (UKCS) – the ninth most important oil and gas province in the world in production terms – recorded the largest oil production decline of all basins in the BP annual review – approximately 230,000 barrels of oil per day. A year later, the basin also had the highest level of exploration, appraisal, and development activities (as measured by number of well starts) since 1998. According to a recent Wood Mackenzie benchmarking analysis[25] of oil and gas basins, UKCS ranked 47th out of 58 in both development costs and finding costs and 42nd out of 58 in terms of commercial success with a chance of 10%. Yet, it ranked 5th out of 63 in terms of exploration wells drilled from 1994 to 2003 and 2nd in terms of expenditure. How did the fiscal regime evolve to maintain such a level of interest – "against all odds"? The answer is by shaping rather than directing the market in line with global conditions, by learning from and reacting to companies' responses and investment alternatives, and by considering changing use and conditions of the North Sea infrastructure.[26]

(1) Early gas field developments delivering to state-owned British Gas had limited profit potential and were charged 12.5% of the wellhead value of gross production as royalty and 52% on companies' taxable profits as corporate tax. In the wake of the 1973 oil price hike, however, it became apparent that this simple fiscal regime would not capture economic rent and achieve a government "take" comparable to that being obtained by other oil-producing countries. Hence the Oil Taxation Act 1975 introduced (a) a 45% Petroleum Revenue Tax (PRT) charged on profits arising from individual oil fields (rather than on all fields owned by a particular company), (b) various methods for tax relief, particularly in view of smaller and marginal fields, and (c)

a ringfence for corporate tax purposes to avoid erosion of North Sea oil profits through intra-company rent-shifting. "The aim of PRT was to allow each project to recover its costs rapidly, then to tax it hard."[27] In the following years, the PRT rate rose with oil prices to 70% in 1982. In addition, a 20% Supplementary Petroleum Duty was imposed in 1981 and 1982. As a result, a high marginal tax rate (ca. 90%) discouraged investment, despite record oil prices of over $80.

(2) Reacting to this, the PRT regime was modified in 1983, permitting new exploration and appraisal costs to be set against existing field profits, and doubling oil allowances for new oil fields outside the Southern Basin of the North Sea. In addition, royalty was abolished for new developments, and a new Oil Taxation Act dealt with fiscal considerations arising from the development of infrastructure in the North Sea and the growth in the sharing of assets – for example pipelines – between two or more oil fields. Although briefly muted by the 1986 price collapse, the 1983 tax-relief pushed up exploration and appraisal activity and led to an average of 95 wells per annum between 1984 and 1992. During the same time the rate of corporate tax fell from 52% to 35%. Still, the differential treatment of fields invited gambling. Companies began to shift income into fields that were not liable for PRT and expenses into those that were. In addition, declining average field sizes reduced the number of fields expected to pay PRT.

(3) Henceforth, fiscal changes in 1993 cut the PRT rate from 75% to 50% for existing oil and gas fields, abolished the tax for fields given development consent after March 15, 1993, and introduced the opportunity to set 100% of expenditure on new exploration and appraisal against profits of existing fields. The industry responded by bringing some 95 projects to approval by 1997, representing over 4 billion boe sanctioned reserves and accounting for 44% of year 2000 production (2 million boe per day). For the UK government, cutting the PRT rate on old fields and abolishing PRT and its reliefs on new ones effectively *raised* the tax yield as it eliminated costly cross-field PRT reliefs.

(4) In 2002, the Labour government introduced a major tax reform intended to balance the promotion of investments and profiting from the use of significant national resources. A new element, a 10% supplementary surcharge, with no deduction opportunities, was added to the 30% corporate tax rate for oil and gas producers in the UK and on the UKCS. In addition, royalty, which applied to older fields, was abolished and a 100% immediate allowance

for most capital investment in the North Sea against corporate tax and supplementary charge was introduced. Although the de facto increase of corporate tax to 40% dampened exploration and appraisal activity, repeated government commitments to stability and rising oil prices increased exploration activity in 2004 by 40%.

(5) In 2005, the UK Department of Trade and Industry (DTI) estimates that some 30 billion boe have been produced, but that the remaining reserves base was in the range of 22–31 billion boe, of which around half had yet to be found. Concerned about majors and North American "super-independents" flocking to West Africa or deepwater Gulf of Mexico, the UK government had earlier launched a number of initiatives to prop up mid-size independents. Promote, a new discounted licensing program, gives companies two years to assess the value of a field before committing to its production. As a result of this, a record high of 27 new entrants to the North Sea participated during the 2003 licensing round. The Fallow Initiative seeks to rejuvenate activity in dormant fields by allowing third parties to propose technical ideas that would stimulate new exploration and development. Finally, a number of tax changes were introduced to encourage new entrants and increased production, such as a direct tax allowance for 100% of new investment. Yet, sustained high oil prices tightened the North Sea assets market during 2004–05, making it close to impossible for new, small oil companies to secure initial stakes. At current oil prices there is simply no reason for the majors to sell, as they are focusing on development and tidying up portfolios around the edges. The DTI stands by.

In sum, the UKCS's fiscal regime has not only been successful in encouraging and directing entry and investments into a maturing basin but, in view of the level of commercial risk and overall cost, it has delivered a disproportionately high level of exploration activity. The fiscal regime evolved to grow both profit and tax return. As a result, the UK government has generated a substantial amount of tax revenue, an estimated £5.2 billion in 2004 alone. Future UK primary energy demand is projected to grow from 233 million tons of oil equivalent (gas 42%; oil 32%) in 2004 to 255 million tons of oil equivalent (gas 48%; oil 37%) in 2020. Under these conditions, accepting maturity as a fact and as a result decommissioning around 40% of assets until 2020 would lead to production being cut to 0.5 million boe per day and 90% of demand being covered by imports at an estimated cost of £25 billion. Conversely, delaying decommissioning by ten years, and producing 1.5–2 million boe per day would meet 50% of total UK oil and gas needs. Extending North Sea production however not only reduces import dependence and costs but also generates substantial tax revenue. The difference between the two scenarios in cumulative tax receipts at 2004 oil prices would amount to £60 billion by 2020. This constitutes a clear win-win – are we seeing this replicated elsewhere?

No Common trend in fiscal regimes

A recent global update of changes in fiscal regimes[28] found that Azerbaijan, Georgia, Ukraine, Uzbekistan, Brunei, Pakistan, Cameroon and Senegal had all switched to production-sharing regimes that offered somewhat better investment and performance terms than previous laws. In addition, more than 20 countries have cut tax rates and excess profits taxes, and are now offering tax holidays and customs exemptions. An equal number of countries improved E&P investment incentives by reducing or completely eliminating state participation, abolishing or reducing royalty rates, or offering preferential treatment of high-cost projects, such as deep water. However, some countries did exactly the reverse. In 2004,[29] Venezuela unilaterally increased its royalty on contracts to develop and produce heavy oil from the Orinoco region from 1% to 16.67%. And the Bolivian draft hydrocarbon law provides for mandatory conversion of existing E&P agreements into service contracts that would strip contractors of their share of production. Likewise, in April 2003, the Russian government stalled international E&P investment by not signing off on 30 already negotiated production-sharing agreements (PSAs) and two months later issued a law, which rendered new PSAs virtually inaccessible to investors. By auctioning off most of YUKOS' assets to a shell company and then transferring these to state-owned Rosneft, the Russian government has effectively moved towards indirect nationalization of its oil and natural gas resources. Clearly, in line with the argument presented above, the Kremlin seems convinced that a fiscal and regulatory system will not suffice to protect the country against the risk of being abused. But is a NOC likely to do this? In what circumstances can a NOC be presumed to be efficient? Given these conditions, can it play a political function? What is the role of a NOC in international markets?

Implications for NOCs

Internalizing the otherwise outsourceable exploitation of a country's oil and gas resource – entirely or partly – in the hands of a single state-owned, state-run enterprise raises efficiency concerns. Internal benchmarking exercises and incentive plans may be unable to substitute for missing market references. Where market principles apply, NOCs are unlikely to pursue non-economic political goals such as employment generation – in general or for particular political allies – the subsidization of domestic oil prices, or investment planning in line with governmental budgetary requirements rather than pure business needs. In addition, the reliance on a nationalized oil producer may distort competition in factors markets, negatively affect the country's non-oil economy, and thereby deepening the dependence on hydrocarbon exports and volatile oil prices. Their international impact is equally questionable. State-financed NOCs are often thought to worry less about environmental and human rights campaigners and may over-bid on assets to assure energy security. Difficult to benchmark domestically, they appear – as a business model – to represent a fall-back into earlier mercantile times and a straightforward contradiction of internationally accepted state-aid rules. And yet, understanding NOCs is vital to assessing the governance of oil and gas resources and must be the focal point of future energy related research.

Summary and conclusions

Observers of current upstream oil and gas developments often equate a change in a company's strategy with a fundamental shift in its underlying business model. This chapter described types of upstream business models and offered a cursory assessment of the strategies pursued by BP, Spinnaker, Anadarko, Baker Energy, Petrofac, Halliburton and Schlumberger, as well as the UK government in managing the UKCS fiscal regime. There seems to be no new or need for new business model. In fact, seen from a conventional international product/industry life-cycle perspective,[30] players appear to develop their given models in line with fairly predictable patterns of industry evolution. While limited opportunities for risk spreading and efficient productive specialization triggered the vertical and horizontal integration of early oil and gas producers, ensuing changes in market conditions, play characteristics and political contexts caused the functional specialization of independent players and technology providers and the operational involvement of international resource owners. In each case, market definition and operational involvement reflect the specific economics of the business as well as coordination and organizational requirements. Currently, the evolution of the LNG sub-sector follows the same pattern, albeit in a sped up fashion. While global majors again provide the initial capital, coordination function and bridge to end-consumers, markets are already deepening and their impact is lessened.

Notes

1. For a review of rather diverse perspectives, see P. Ghemawat and P. Botticelli, "Competition and Business Strategy in Historical Perspective," *Harvard Business School*, Teaching Note 9–798–010, 1997; J. Magretta, "Why Business Models Matter," *Harvard Business Review*, May 2002, pp. 86–92; A.O. Krueger (ed.), *Economic Policy Reform*, University of Chicago Press, 2000; W. Adams and J. Brock (eds), *The Structure of American Industry*, Prentice Hall, 1995; G.M. Meyer and J.E. Stiglitz, *Frontiers of Development Economics*, Oxford University Press, 2001.
2. I would like to thank Richard Doidge (President, Maersk Oil America, Inc.) for discussing these with me.
3. G. Bouckaert and C. Pollitt, *Public Management Reform: A Comparative Analysis*, Oxford University Press, 2000.
4. The average extraction cost in the North Sea for a barrel of oil rose 42% from 2000 to 2005. See I.J.H. Bozon, "What's next for Big Oil?," *McKinsey Quarterly*, Issue 2, 2005, pp.94–105.
5. For example, in 2000 BP was put under a lot of pressure from Christian Aid and others in connection with their investments in PetroChina, whose holding company operates in Sudan.
6. 3i Oil and Gas, *The Prospects for North Sea Oil and Gas: Challenges and Opportunity in a Maturing Province*, Aberdeen, 2005.
7. For example, Shell's Smart Fields program, a series of R&D partnerships with Schlumberger Information Solutions, Invensys, IBM, Intelligent Agent Corporation, Science Applications International Corporation and Microsoft. Signed in October 2004, the goal was to accelerate the development and deployment of next-generation digital oils and gas-field technology. See "Shell Smartens up with Technology Partnerships," *Aberdeen Press and Journal*, October 4, 2004.
8. Citigroup Analyst Report. BP plc (BP), October 26, 2005.
9. R. Casadesus-Masanell, D.J. Hanson and F. Reinhardt, "BP and the Consolidation of the Oil Industry: 1998–2002," *Harvard Business School*, Case 9–702–012, 2003.
10. Global majors and host countries typically take on 50% and 30% of the LNG infrastructure investments respectively, and buyers and trading houses contribute the rest.
11. Case interview – 38. Casadesus-Masanell et al., "BP and the Consolidation of the Oil Industry: 1998–2002."
12. Ibid.
13. For background information see IPAA, "Profile of Independent Producers," July 1998, USA; W. Furlow and M. DeLuca, "Shelf Mini-Boom Triggered as Majors Disgorge Hundreds of Leases," *Offshore*, Vol. 60, Issue 7, 2000; G.C. Reid and T.M. Stewart, "Independents Abroad: The Pursuit of Expansion by Independent Oil Companies into Non-Traditional Petroleum Countries," Working Paper, University of St. Andrews, 2005; T.M. Stewart, "Indigenous Wildcatters, Clandestine Prospectors of the Petroleum World: An Exploratory Study of British Independent Exploration and Production Companies," Thesis presented at the School of Economics and Finance, University of St. Andrews, 2005.

14. In 1998, 17% of an IPAA survey of US independents reported that they operated internationally, with more than 25% planning to pursue international opportunities within the next five years – their geographical focus was Canada and South America. IPPA, "Profile of Independent Producers."
15. In the UK, the 13 independents making up the Association of British Independent Oil Exploration Companies (BRINDEX) were involved in 184 UK Continental Shelf licenses (48 as operator), in 22 UK onshore/seaward licenses (18 as operator) and had petroleum interests in 30 countries overseas. While British independents, irrespective of size, were interested in internationalization, only large-scale North American operators (market cap > £1,000 million) would do so. See Reid and Stewart, "Independents Abroad."
16. Proved developed producing (PDP) reserves are the most likely to be recovered while proved undeveloped (PUD) are the least likely to be recovered. In addition to proved reserves, oil and gas companies record probable, possible and potential reserves.
17. Reid and Stewart, "Independents Abroad."
18. Based on the analysis presented by Claire S. Farley, CEO of advisory firm Randall & Dewey, speaking at the Deloitte Oil and Gas conference in Houston, September 16 2003.
19. Or $22.65/boe based on Spinnaker's proven and probable reserves.
20. N. Nohria and K. Palepu, "Leading Anadarko," *Harvard Business School*, Case 9–406–014.
21. Ibid.
22. Ibid.
23. J. Kammerzell, "Independent Operator Provides Complete Turnkey Abandonment Solution in GoM," *Offshore*, August 2005, pp. 90–2.
24. Merrill Lynch Analyst Report, *Schlumberger Ltd.*, October 18, 2005, pp. 5–6.
25. Quoted from R. Hall, "Living with Fiscal Instability – the UK Experience," presentation given at the IQPC conference "Stabilizing Fiscal Terms in Upstream Oil and Gas," London, August 17–18, 2005.
26. For a detailed reference, see M. Earp, *Review of UK Government Tax Policies as Applied to North Sea Production of Oil and Gas*, Energy Group, UK Department of Trade and Industry, March 2003.
27. Ibid.
28. "Major Shifts in Global Oil and Gas Industry Spin Off More International Opportunities for Independent Producers," *Oil and Gas Investor*, March 1, 2005.
29. Economist Intelligence Unit Ltd., *Venezuela Risk: Risk Overview – Executive Briefing*, London, December 2004.
30. For the original writing, see R. Vernon, "International Investment and International Trade in the Product Life Cycle," *Quarterly Journal of Economics*, Vol. 80, May 1966, pp. 190–207.

8
The Future of Innovation in the Energy Sector – Is it Doomed?

Aruna Subramanian

Despite continuing healthy trends in oil prices, research and development (R&D) expenditure by the oil giants and governments has been shrinking over the past decade. With the growth of partnership arrangements between oil companies and service providers, the international service companies have accelerated their research programs recognizing that it is in their own best interests to do so. Current industry contractual practices however, leave little incentive for the service companies to invest in innovations that do not yield short-term benefits to help recoup their investment and broad implementation of the technology. Market forces drive the majors to focus on short-term results, not long-term R&D investments. Industry R&D focus, therefore, will gradually shift away from long-term technology strategies.

Who then sustains disruptive innovations? The small and medium enterprises (SMEs) remain the major source for concept generation but the capital-intensive nature of the industry, combined with the slow uptake of new technologies, provides a formidable barrier to success. Is innovation in the oil industry then doomed?

Introduction

The oil and gas exploration and production (E&P) industry has traditionally been slower than other industries to accept innovation. Now more than ever before it is evident that novel technologies are needed to improve the recovery factor of identified reserves to meet ever-increasing energy demands. However, industry statistics show that it takes over 30 years for an idea to progress from its conceptual stage to commercial status. In the past decade, the oil majors have consistently reduced their R&D expenditure and have relied on service providers to come up with innovative solutions. The international service providers, driven by ineffective risk-reward sharing contractual arrangements and a push by the majors to commoditize new technologies as rapidly as possible, do not have adequate incentives to come up with disruptive inventions. The long deployment process and complicated equity partnerships make it difficult for the small entrepreneur to survive. So does innovation stand a chance of survival in the energy industry? And who then is responsible for it?

Why do we need innovation?

It is no secret that the global demand for oil has been on a steady increase. Despite incredible progress in the realm of renewable energy sources, a report by the Energy Information Administration (EIA) of the US Department of Energy (DOE) projects that over the next two decades the bulk of energy demand will continue to be filled by conventional hydrocarbon sources.[1] Just under 14% of the estimated 134 quadrillion Btu (British thermal units) of energy consumption in the United States in 2030 will be fulfilled by renewable, nuclear and other energy sources. The report further indicates that the transportation sector leads the drive for energy use in 2030 followed by the residential, commercial and industrial sectors.

Clearly, oil is a non-renewable resource and debates about when the global "Hubbert's Peak"[2] will be reached are once again clouding the industry's future outlook. But in spite of record consumption rates, EIA estimates on the ultimate recoverable resource base have consistently grown over the past few decades. The 2006 *Annual Energy Outlook* report[3] indicates that the global proven crude reserve estimates were about 700 billion barrels in 1985, increased to about 1,000 billion barrels in 1995 and grew even further to about 1,300 billion barrels in 2005.

What makes the estimates go up continuously is a combination of economics and innovation. Reserves are constantly revised with new discoveries, changes in prices and technological advances. These revisions invariably add to the reserve base. A few decades ago, the average oil recovery rate from reservoirs was 20%; thanks to remarkable advances in technology, this has risen to about 35%[4] today. But despite this improvement, two-thirds of the oil known to exist in reservoirs is still abandoned as uneconomic, leaving room for disruptive discoveries or innovations to lift recovery rates and push Hubbert's peak even further into the future.

Experts had predicted that fields in the British North Sea would reach their maximum output by 1990. In fact, they have only just peaked with technological advances such as deepwater and horizontal drilling vastly improving the recovery factor, making previously commercially unattractive fields viable.

There are several examples that have proven reserve depletion forecasts wrong. Russia demonstrated a dramatic increase in production when the markets opened to private investment causing a huge surge in technology in-flow and entrepreneurial growth. Brazil's Petrobras reserves have increased by 3.4 billion boe (barrels of oil equivalent), a growth of 35%, over the last three years and 80% of that is in deep and ultra-deep waters. This would not have been possible without immense focus on new technologies and the ability to deploy them effectively.[5] Saudi Arabia' Minister of Petroleum and Mineral Resources, Ali Al-Naimi, is confident that Saudi Arabia's current estimated reserves of 1.2 trillion barrels is very conservative and new technologies will help discover new resources and extract even more oil from existing reserves.[6] The list goes on.

Even Dr. Hubbert did not get it quite right. His forecast for American production ignored the vast quantities of oil in the deep waters of the Gulf of Mexico. The industry is exploring under water at depths that were unimaginable 20 years ago. The US Minerals Management Service quotes discoveries at depths of 7,000 feet in the Gulf of Mexico![7]

This may seem an unfair criticism of Hubbert, as he had no way of predicting the wave of offshore exploration, drilling and production technologies that have become available in the past decade. But that is the point: tomorrow's innovations can prove today's experts wrong.

The good news is that there is plenty of oil still available and the key challenge is to get it to the market. Halliburton's David Lesar summarizes, "Today's innocuous technologies could prove tomorrow's breakthroughs, as long as the industry continues to encourage innovation."[8]

What has happened to innovation?

It is a known fact that the E&P industry invests less in R&D than most other technology dependent industries. A 2001 McKinsey report[9] indicated that compared to an R&D investment of 13.6% of revenues in the IT industry, 11% of revenues in the health sector and 5.3% and 2.4% of revenues in the electronics and telecommunications industries respectively, the oil and gas industry spends a mere 0.9% of revenues on R&D!

Data from the US Patent and Trademark Office shows a much slower growth in the number of patents filed in the E&P industry in the last three decades compared to other industrial sectors such as industrial automation, construction and automobiles. This is also indicative of poor patent protection and fast "duplication" of solutions in the oil and gas industry.

Six major groups play a part in the development of innovative ideas in the E&P industry from concept to significant market penetration, and their levels of involvement vary in each development phase (Figure 8.1).

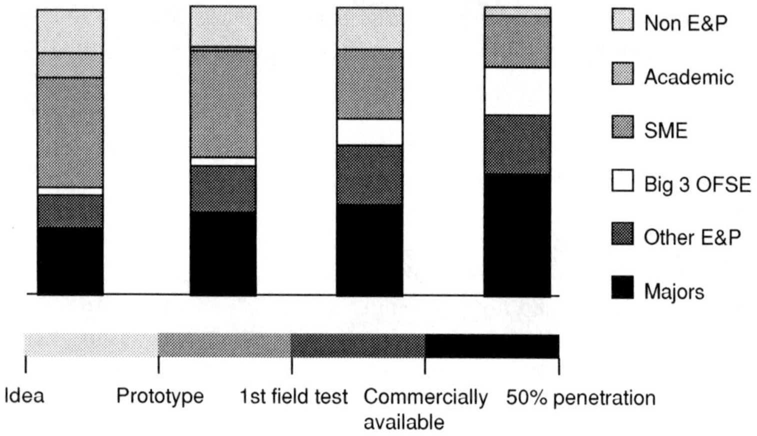

Source: Adapted from McKinsey (see Note 9 at end of this chapter)

Figure 8.1 Role of Various Players through the Development Process

The oil majors typically retain in-house R&D capabilities for core competencies and strategic focus areas where stringent Intellectual Property (IP) ownership is required. However, most of them are increasingly moving to an autonomous asset-based business model. Multi-owner joint-venture decision structures and cannibalization risks slow down effective technology deployment. The operating units, driven by production metrics, are focused on short-term commercial goals and do not have adequate incentives to take risks on production through new technologies. They are therefore risk-averse, leading to a slowdown in intellectual creativity and energy around disruptive innovation. The relative unattractiveness of the industry in an increasingly competitive human resource pool has simply expedited the decline of new technology development.

The short term memory and interest of the market does not attribute sufficient value to long-term investments, reducing the incentive for oil companies to focus on "out-of-the-box" R&D. As a result, the oil majors have demonstrated a consistent decline in their R&D spending[10] and the onus of technical innovation has been outsourced to the international service providers.

The international service companies have taken on the responsibility to be the primary technology providers in the industry but even their increased investments have failed to stem the overall downward trend of technology development. The impact is evident from the number of published patents by the major service companies versus the number submitted by the oil majors over the past decade[11] – the former showing a significant upward trend compared to the latter. In 2001, the service companies recorded over 450 patent awards while the super-majors came in just under 150 awards, indicating a large inequity in the technology focus and ownership between the two groups.

The reason is simple. The international service providers have long since lost the incentive to innovate. In the current contracting environment, the risk and investment involved in developing truly novel technologies is not sufficiently shared. So the service companies increasingly tend to broaden their technology portfolio through acquisition of smaller, more adventurous players, as is evident from the mergers and acquisition activity in the industry. It is far more attractive to take an active role once the success of the technology is proven and commercially realizable, rather than risk large capital investments in long-term, unproven solutions.

This shift of knowledge in the industry from the oil majors to the service providers has resulted in an information asymmetry leading to the devaluation of innovation, the seller valuing it higher due to higher understanding of, and investment in, the technology and the buyer devaluing it to accommodate cost drivers. The imbalance in technical focus has resulted in a bizarre situation in which the "supplier" develops and values a solution more than its "beneficiary."

The current operating model of most oil majors further complicates this exchange of technology. As a poll at a recent industry conference indicated, oil and gas operators demonstrate a lower tolerance for risk-taking in scenarios of high reward and high risk even if the calculated net present value (NPV) is the same.[12]

The niche-focused E&P players, such as in coal bed methane, have maintained their areas of expertise partly due to lower levels of interference by the majors and lack of interest on the part of service companies. They continue to innovate, albeit concentrating on their specific technology applications. But they do not have the critical mass to make a macro impact on the industry.

Industry research consortia have typically been joint efforts by oil majors, service companies and academic institutions. Joint research projects have definitely been instrumental in raising general understanding of new technological advances. But there is a reluctance to share confidential information and IP ownership is unclear as concepts with yet "undefined" value are researched. For instance, though there may be consortia collaborating on novel reservoir surveillance techniques, better and faster identification of new reservoirs is a primary competitive edge and the core basis of the business for any oil company so knowledge sharing is guarded, at best. In spite of their good intentions, therefore, these consortia only infrequently result in successful business enterprise due to their lack of focus, commercial discipline and accountability.

The next two groups – the academic and non-E&P institutions – are often thwarted by the "not invented here" attitude in the traditionally conservative oil and gas industry. Concepts generated "outside" often do not cross over into E&P applications successfully. Furthermore, without industry support, these bodies do not have the necessary independent financial strength to continue research for the long uptake periods characteristic of this industry.

The small and medium enterprise (SME) population is by far the richest source of disruptive ideas and novel technologies in the industry, responsible for approximately 40% of concept generation. But less than half of them succeed; the small entrepreneurs are faced with formidable challenges. Compared to other industries, E&P has traditionally been very conservative in technology uptake, taking over 30 years to get an idea to commercialization. This compares to less than 16 years for the adoption of ADSL broadband telecommunications and around 12 years in the medicine industry. Survival for over 20 years before first commercial deployment requires significant financial staying power.

To raise entry barriers even higher, investments have been very cyclical in the oil industry, its volatility mirroring the trends in oil prices. Long return times and high capital requirements deter venture capitalists. Therefore, though there are several seed projects, there are fewer investments to take the concept further along the path to commercialization. Few investors understand advanced technology and are willing to risk dependence on a few key customers. Statistics show that there is a shortage of funding compared to other industries. In 1999 in Europe, compared to US$2.9 billion of available venture capital for communications, US$639 million for construction and US$518 billion for electronics, only US$202 million dollars of venture capital was available for the energy industry. This equates to 0.02% of the market capital value – an extremely conservative amount that indicates little faith in the ability of new technologies to survive to provide adequate returns on investment.

In spite of all these obstacles, if a hardy entrepreneur does develop a disruptive product, the infamous "technology adoption chasm"[13] comes to play. Early investors may see the idea through to proof of concept but the inherent risk-averse nature of the industry yields rare test opportunities leaving the entrepreneur unable to prove the technology value proposition to engage the next level of financial investors. The traditional private equity firms and venture capitalists do not have the industry resources to secure these opportunities hence the chasm is a death trap for many an innovation.

The very high level of inertia in the industry, limited funding sources, the widespread presence and contractual strength of larger service companies and the high cost of capital pose formidable market entry barriers to the most determined entrepreneur. This is summarized in Figure 8.2.

Does this mean that we, as an industry, are condemning our richest mine of entrepreneurship to inevitable doom? How do we then promote, or at least sustain, a constant in-flow of novel ideas and thinking in the E&P arena?

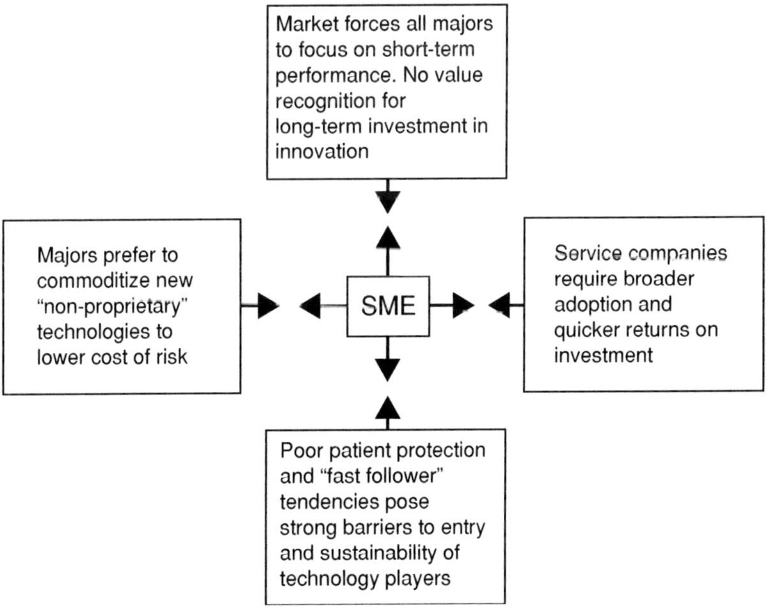

Figure 8.2 The Opposing Forces to Entrepreneurial SME Activity in the Oil and Gas Industry

Who then is responsible?

Disruptive innovation will die in the oil and gas industry unless the majors actively show commitment to promoting new technologies. Unfocused industry consortia do little more than maintain the illusion of collaboration and progress – lack of accountability and conflict in knowledge ownership leaves the effort akin to long drawn out and unfocused laboratory research. While the service industry has certainly picked up the responsibility pressed upon it, there are inherent gaps in its ability to fully understand the needs of the industry, due partly to the nature of the current contractual relationships and partly to the confidential nature of exploration strategies. This lack of symmetry in drivers and recognized value is an opportunity to create a new approach to technology management in the industry.

It is critical that the oil companies take on the responsibility of promoting technological creativity and out-of-the-box thinking for the long-term sustainability of the industry. If the last decade can be used as reference, technology certainly will help lengthen the plateau of oil and gas supply through more economic recovery of current assets and indeed through novel methods of discovering and exploiting new and challenging frontiers. The oil companies are accountable to the market and to their shareholders to deliver energy faster, cheaper and more effectively. New technology is the key; in fact, the key to the very survival of these companies.

A focused industry effort is necessary to promote modern economic valuation and risk assessment methods for decisions made by the operating units. The decision whether to use new solutions must be based on objective assessment of the value added rather than a human tendency just to meet set production goals and avoid risk that could significantly improve recoveries. Access to low-risk assets must be facilitated to ease the field trial process for innovators so that they can demonstrate their value proposition and progress further on the development staircase. Leaving a technology untested on the shelf benefits no one.

Corporate venture capital from the oil majors is needed to keep small technology enterprises from the death trap between initial field tests and established market penetration. The "small company" environment is often more conducive to developing and commercializing certain technologies efficiently and effectively without cumbersome corporate structures. Strong patent protection and accountability create focus and more effective collaboration between the majors and the innovators to ensure survival of the technology, without compromising returns on investments.

Enventure Global Technologies (Enventure)[14] is evidence that this form of collaboration works. This joint venture company was formed in December 1998 between the corporate venturing arm of Royal Dutch Shell – Shell Technology Ventures (STV) – and Halliburton Energy Services (HES) to commercialize the concept of solid expandable tubulars that originated in Shell's research laboratories in the early 1990s. At the end of 1997, STV and HES began exploring the possibility of forming a new limited liability company whose initial focus would be the development and commercialization of expandable tubular technology. The new company would be funded and managed equally by the two partners. On December 9, 1998, Enventure Global Technology L.L.C. was born.

In May 1999, Enventure announced its three new product lines at the Offshore Technology Conference in Houston, USA. In late November of that same year, the company successfully and safely installed the first open hole liner (OHL™) in a Chevron well located in the West Cameron 17 Field in the Gulf of Mexico. Expandable tubular systems had become a commercial reality. Since then, Enventure has revolutionized drilling, making it possible to have larger boreholes than was previously feasible and is now the market leader in expandable solutions, setting and breaking its own records. As of March 2006, it has completed over 427 installations for 82 customers in 20 countries. This groundbreaking achievement was possible in less than 15 years from conceptualization, half the average time it has historically taken in the industry.

This example confirms that focus and collaboration can and will work in keeping innovation alive. However, it was not just a matter of chance. The newly set up Enventure has its own management team and was allowed to flourish independently. Though powerful industry players were needed to provide the capital, technical expertise and field-test opportunities required for the successful uptake of this technology, there was a risk that either or both of the strong company cultures would dominate the start-up. An independent management team with its own IP ownership and business goals helped retain the entrepreneurial efficiency and focus. Industry experience meant that the two key investors understood what it took to succeed in the E&P industry – a vital differentiator from traditional financial investors.

Enventure is a success story driven by two major players but there are several other examples of successful collaborations between a major industry player and innovative small companies. vMonitor[15] – a small start-up company supported by STV and another major service company, specializes in remote monitoring on a global basis. Sensornet, a UK based fiber optic company, is also supported by STV and well on its way to realizing the full potential of its technology in the oil and gas industry. And the list goes on.

Conclusions: what's next?

Focused and disciplined collaboration between the major industry players such as in the case of Enventure will be critical to the successful development of disruptive technologies to meet the energy needs of tomorrow. Several areas are open to renewed focus and efforts. There is significant room for new artificial lift, enhanced production and efficient drilling technologies to effect a step change in recovery factors. New surveillance technologies are needed to tap into yet undiscovered reserves. Renewable energy developments such as wind, wave, tidal and hydrogen technologies are still largely based on unsustainable public finance. The industry needs to take up the onus of clean technologies to complement the strong presence of existing fossil fuels and develop them in a sustainable manner. A plethora of carbon solutions needs strong sponsors to maintain environmental standards. Most of the innovations in these domains require SMEs to partner with strong industry players, while retaining their focus and independence. Their success may allow Hubbert's peak to be moved another step further away.

Notes

1. Energy Information Administration, *February 2006 – Annual Energy Outlook 2006 with Projections to 2030*, Report no.: DOE/EIA-0383 (2006).
2. The legendary and late Dr. M. King Hubbert, geophysicist, is regarded a world authority on the estimation of energy resources and on the prediction of their patterns of discovery and depletion. He shot to fame with his startling prediction, first made public in 1949, that the fossil fuel era would be of very short duration. His prediction in 1956 that US oil production would peak in about 1970 and decline thereafter was ridiculed at the time but his analysis has since proved to be remarkably accurate. See <www.hubbertpeak.com/hubbert/>.
3. Energy Information Administration – International Petroleum (Oil) Reserves and Resources Tables and Reports: January 1, 1980 – January 1, 2006 Estimates.
4. IEA, *Oil And Gas Summary Report 2005*, <www.iea.org/textbase/npsum/oil_gasSUM.pdf>.
5. Carlos H.L. Bruhn, *How Much Oil and Gas from Deep Water? The Experience of Brazil*, Rio de Janeiro: Petrobras, 2005.
6. *Arab News*, "Saudi Oil Is Secure and Plentiful, Say Officials," April 29, 2004.

7. US Minerals Management Service – Gulf of Mexico Region Offshore Information <www.gomr.mms.gov/homepg/offshore/deepwatr/deeptbl2.html>.
8. *The Economist*, "The Bottomless Beer Mug," April 28, 2005.
9. McKinsey, "A New Regime for Innovation and Technology Management in the E&P Industry," October 2001.
10. Analysis of R&D Investments from Annual Reports by Cambridge Energy Research Associates (CERA) – from the McKinsey report: "A New Regime for Innovation and Technology Management in the E&P Industry," October 2001.
11. Derwent WPI, Global patent search on Upstream Code; Charles River Associates, Published patents (1991–2002).
12. R. Rodriguez, *Accelerating Technology Acceptance (Risk and Return)*, 2005 Offshore Technology Conference, Houston, Texas, May 2–5, 2005.
13. G. Moore, *Crossing the Chasm: Marketing and Selling Technology Products to Mainstream Customers*, HarperCollins, 1991.
14. Enventure Global Technologies website – <www.enventuregt.com/index.htm>.
15. Shell Technology Ventures, <www.shelltechnologyventures.com>.

9

Reviewing Areas of Engagement: Potential and Pitfalls

Olav Ljosne

The chapter starts with an analysis of potential societal development, focusing on three global scenarios developed by Shell:

- the civil-society-centric world;
- the market-centric world; and,
- the state-centric world.

It then looks at who has influence in each scenario and also at the potential risks that each entails for business and for oil companies.

Moving from the future back to the present, the chapter goes on to consider the risks and issues faced by companies in today's society. Further, it investigates whether business really understands the society in which it operates and in which it seeks a license to operate further. The last two sections seek to answer the following questions:

- Do the leaders of the oil and gas business understand its short- and long-term impact, and its implications for society and for the business itself?
- How does one manage the risks and issues that the oil business faces?

Introduction

Do we really understand the world we operate in? Or does our understanding stretch only to where we feel most comfortable?

Over the coming decades, energy companies will be forced out of their comfort zone by a range of factors: not least, the challenge of meeting the 60% increase in global energy demand forecast by the International Energy Agency (IEA) between 2004 and 2030.

Given that 85% of that increase will, according to the IEA, be met by fossil fuels – because other sources simply cannot be brought on line fast enough – coal, oil and gas (including unconventionals, which by 2030 could represent 10 million barrels per day) will remain firmly in the center of the radar screen for some time to come. Shell scenario work shows that, with strong government support, renewables could grow by 10% per year; but even so they would still constitute less than 10% of total energy supply by 2020, up from 1% today.

Alternatives to fossil fuels have the potential to be significant after 2025. However, economic growth and development in the intervening decades will rely on the current dominant fuels; and major energy-using countries will increase their dependence on energy imports.

Where will those imports come from? The earth's "easy" oil and gas resources have largely been tapped. Now the industry needs to take on the challenge of finding and developing new, more difficult resources while at the same time making the most of existing assets. Yet many of the major resources are found in regions that could be described as "difficult" for political reasons, or that present environmental or geographical challenges – the Arctic, deepwater in the Gulf of Mexico, Siberia, sub-Saharan Africa, CIS, etc.

In certain parts of the world, making the most of known assets means applying innovative technologies to get more out of the ground in an environment – economic, political and societal – that is stable and predictable. In other regions, quite a different situation prevails: the challenge is much less on the technology front, and more to do with the stability and safety of the environment in which the energy industry seeks to operate. Companies need to develop ever more robust ways of assessing and analyzing the upside and downside balance in a given situation, so that decisions that will impact a wide range of stakeholders for many years to come can be made with confidence.

Scenarios for a complex world

The world, as a whole, is not an easy place to understand. It is complex – and, many say, growing steadily more complex. Demographic shifts, poverty, lack of education and health services, the rapid emergence of developing nations and growth in global communications are all helping to fill the landscape of the future with challenges for energy companies: not just in terms of meeting the forecast rise in energy demand, but in doing so in a manner that is sustainable in every sense – economically, technologically, socially and environmentally.

Advances in science and technology are being made to help deal with the challenges, but these require investment – as much as $16 trillion between now and 2030, according to IEA estimates – and effective management to deliver what is needed by the time it's needed. Meanwhile, global phenomena such as the "war on terror" and natural catastrophes such as Hurricanes Katrina and Rita increase the volatility of the picture, requiring additional flexibility and contingency to be built into business strategy and planning.

For 30 years, Shell has used scenarios to give the top leadership alternative thoughts on where the business environment might go. These scenarios help to address some fundamental needs in a global business that invests money in projects lasting up to 50 years or more. They provide a way of dealing with time scales longer than anyone can make valid predictions for; and a means of identifying critical risks and opportunities in the global business environment, highlighting societal and other uncertainties ahead that might affect our business.

Scenarios are reasonable and credible answers to the "what if?" questions that underpin survival in business. They are not forecasts or projections. They are not predictions of what's to come. Nor are they preferences, views of how one company would like the world to evolve. Instead, they provide alternative views and thoughts on the future. In the case of the Shell Global Scenarios, that future is first and foremost relevant to business – the scenarios are developed to help us run our business better.

One of the most important characteristics of the scenarios is the fact that they challenge assumptions: nothing is a "given" or "obvious." Every premise is examined carefully: "How do we know that?" "Is there an alternative, equally valid, assumption that could be made?" As a result, the Shell Global Scenarios provide a more robust mechanism for testing strategies and for managing risk.

Our latest round of scenario work has produced *Shell Global Scenarios to 2025*. Published in 2005, it examines some trends that are shaping the world and the global business environment. These trends are called "predetermined": they will continue to be dominant and will play out, in one way or another, over the next two decades.

Key predetermined trends include: the increasing role of China and the USA as agents of global change; the explosion in urban populations and migration between countries; and the growing importance of carbon, in relation to climate change and also as a commodity in its own right. All of these have profound implications for business, society and the environment.

How each trend will play out is uncertain. Future developments will be determined by some critical uncertainties, i.e. uncertainties that will have significant impact in shaping the future. For the purposes of the latest *Scenarios*, one uncertainty stands out above all the others. And it has its roots in what is very much a twenty-first-century crisis.

Some people would say that there are two crises – a crisis of security, with the attacks of 9/11 as an example; and a crisis of trust in the marketplace, where Enron provides one of the best-known, but by no means the only, example. In reality, these combine to form a dual crisis of security and trust.

Together they challenge the view represented in previous rounds of scenarios, which envisaged the world trying to move in two directions at the same time: towards more democracy, and towards a more market-friendly, and even market-driven, world. The two goals are not mutually exclusive, but neither are they mutually reinforcing. So, in previous scenarios, we saw a linear tension created: between market forces and the force of community. The reach and power of the nation state, meanwhile, had been steadily eroded by globalization, liberalization and technological developments.

This linear picture no longer holds. The events of recent times have brought government's role as the guardian of security and market integrity back into the picture.

In effect, government has a more popularly accepted mandate to protect the physical and financial wellbeing of both citizens and businesses. Whereas in the 1990s government was often seen as "part of the problem" for business, now the state is once again perceived as part of the solution. Society and businesses expect governments to provide security – security from terrorism, security from international conflict – and also to restore trust in the marketplace, policing the marketplace much more effectively.

How the state responds varies from nation to nation. However, whether they are imprisoning hundreds of suspects without trial, or jailing heads of major oil companies on the grounds of tax fraud, it can be seen that states are increasingly exerting the power they have – through coercion and regulation – to restore security and trust.

Shell Global Scenarios to 2025 represents this move away from a linear model into a newly developed framework, known as the Trilemma Triangle. The apexes of the triangle represent three theoretically possible, but wholly unrealistic, worlds: one centered completely on the state, the second on civil society and the third on the market.

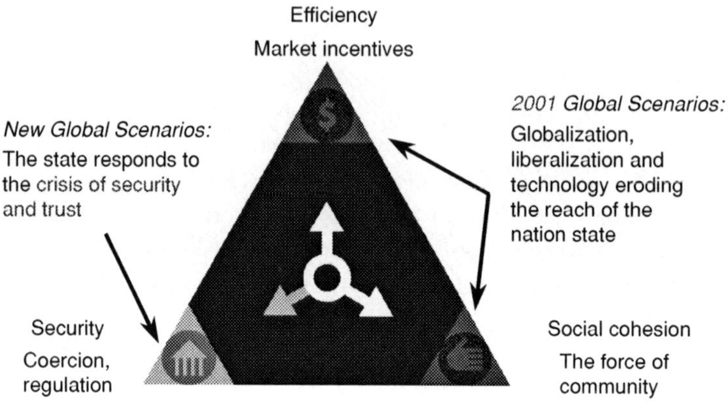

Source: Shell

Figure 9.1 Three Forces Shaping Our Global Business Environment

Such worlds would never exist, because the real world is all about trade-offs, as competing forces – market forces, the force of community, state forces – pull in different directions. For example, the state tries to ensure greater security and trust, but at times the steps it proposes encroach on civil liberties, and/or endanger social cohesion. Similarly, measures to foster the markets' drive towards greater efficiency can sometimes be impeded by the desire not to widen the gap between rich and poor any further.

The points *within* the Trilemma Triangle capture the kind of trade-offs that happen in the real world. The world we live in is shaped by coalition – by political forces, economic forces and civil groups coming into contact with each other and reaching some sort of compromise.

The shaded zones in the triangle in Figure 9.2 represent this kind of complexity in a manageable way. Each one is formed by the overlap of two circles. For example, the zone marked with a "Y" is the overlap of the circle centered on the market incentives apex, and the one centered on the coercion and regulation apex. So this Y-zone describes a world with more globalization, *and* more power of the state, more regulation.

This may seem a little surprising; because so many thinkers tell us that this cannot happen: we are told that "more market means less state." However, we have at least one real-world example of life in the Y-zone: in the USA we have a country that has introduced more tax cuts while increasing its defense budget; one which encourages financial innovation but has also passed the Sarbanes-Oxley Act, which will put the CFO in jail if he does his job badly.

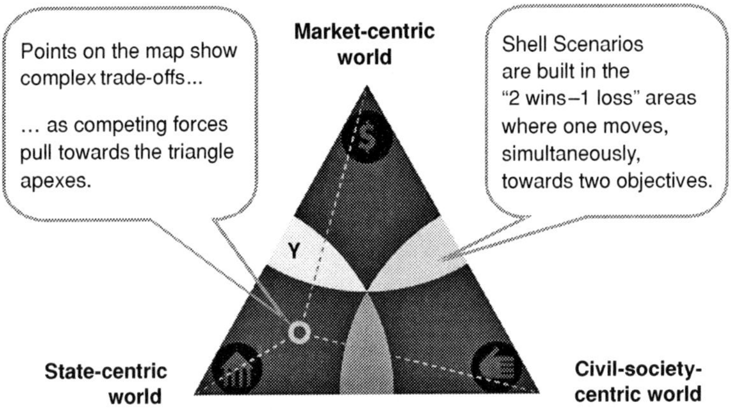

Source: Shell

Figure 9.2 Mapping Competing Forces that Influence Strategies and Behaviors

Thus the real world is about things that are combined, sometimes in surprising ways. The working assumption in *Global Scenarios to 2025* is that, if nothing happened – such as another major terrorist attack – to take the world into a completely different ball park, the three shaded zones in the Trilemma Triangle are where the forces would be expected to find some stability. That is because these zones are where the largest possible coalitions will be satisfied – investor groups, government groups, groups in society. These shaded areas represent the "win-wins."

The three scenarios are designated respectively Low Trust Globalization, Open Doors and Flags, as shown in Figure 9.3.

Low Trust Globalization can be described as a skeptical, "prove it to me" world. The key words are about compliance, about playing by the rules. International companies need to be very aware of the diversity of jurisdictions in which they operate, each with very different demands on business. Heavy compliance requirements in the world of Low Trust Globalization bring high costs and foster scale: markets are dominated by large, vertically integrated companies. Compliance and its cost and scale implications also create large barriers to market entry for small and medium-sized companies.

The second scenario, Open Doors, is – in contrast – a "know me" world, a world of trust: trust in global systems, and in globalization. Like Low Trust Globalization, Open Doors already exists in the reality of business. It is a world where Scandinavians have proved themselves to be role models. These countries exhibit high trust in globalization,

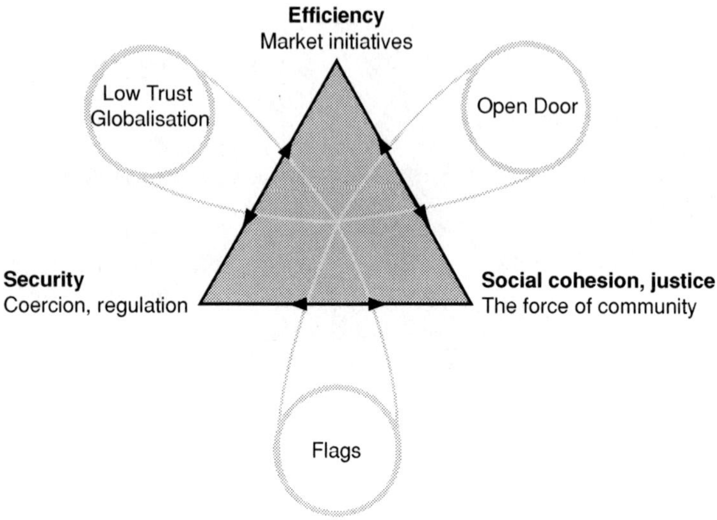

Source: Shell

Figure 9.3 The Shell Global Scenarios

and are very open to it – largely because people within the Scandinavian countries feel protected from globalization's downsides, by their generous and comprehensive social welfare systems. Evidence of their success lies in the fact that they are consistently among the leaders in the annual World Economic Forum competitiveness leagues. In general we can see that the world of Open Doors combines social justice and cohesion with competitive markets.

Finally, the world of Flags is fragmented and polarized – both internationally and domestically. It is a "follow me" world, where people adhere tightly to their codes. In a Flags-type nation, life is often about being against things. People distrust elites, they distrust one another. They want to show their differences from others by stating them clearly and saying "no" to anything that is counter to their position, views or beliefs. Flags is not a world of globalization in reverse, but globalization proceeds much more slowly here than in other scenarios. Opposed groups in society try to capture the state for their own objectives, by one route or another: votes, violence or money.

In each of the scenarios described, it is especially interesting for businesses to note where most influence lies.

In the Low Trust Globalization scenario, regulators act through law enforcement. Bureaucrats and lawyers are very important people. Civil society, represented by non-governmental organizations (NGOs), is not without influence – but it acquires and exercises influence through alliances with business, with state or media organizations (nicknamed BINGOs – business initiated NGOs, or GRINGOs – government related NGOs), and with financial institutions. The media report on corporate performance and corporate scandals, but bear a serious risk to their reputation from any inaccuracies or opinions that cannot be backed up strongly by facts.

Open Doors features regulators as facilitators. They foster and channel the many private initiatives that flourish to certify, assess and promote trust-enhancing norms and practices; business success hinges on maintaining reputation, and such norms and practices are seen as essential to preventing reputation damage. Civil society is also a source of supervision and screening, with NGOs as the voices of endorsement for contributions by companies and government towards shared goals and also towards specific causes embraced by influential groups (e.g. faith-based organizations). The media, meanwhile, play a vital role on their own terms in empowering stakeholders to hold companies, governments, individuals and even NGOs accountable for initiatives they take, or fail to take.

The world of Flags is characterized by political leaders pursuing populist policies and by trade unions and other types of community each claiming the sole allegiance of its members. NGOs act as the voice of public discontent. Media scrutiny focuses on political compromises and connections that distort markets, rather than on the standards of social responsibility that we see emphasized in Open Doors.

Trust and transparency

Shell Global Scenarios to 2025 presents three quite different ways in which the major uncertainty ahead for all of us – encapsulated in the dual crisis of security and trust – may be resolved.

As might be expected, Low Trust Globalization features intrusive security checks and controls, and regulation through rules-based mandatory standards – proof is everything. The higher-trust Open Doors scenario relies on a collaborative, inclusive approach that speaks to the self-interest of businesses through self-governance to preserve reputation. Flags represents, by contrast, a world of isolated communities with heavily policed boundaries; groups seek to protect their values by requiring strict adherence to them, and conforming is the key to being trusted.

Table 9.1 Three Contrasted Resolutions of the Dual Crisis

Scenario	Security	Trust
Low Trust Globalization	Intrusive checks and controls Ad hoc crisis management	Rules-based mandatory standards "Prove it to me" world
Open Doors	Inclusive institutions, alliances, precautionary principle	Principles-based standards, voluntary codes Reputation premium "Know me" world
Flags	Reduced dependency, check points, gated communities	Affiliation, loyalty and community based "Follow me" world

Source: Shell

Much of this emphasis on trust is born of an evolution in global society that has taken place over the last few decades, with the emergence of the "show me" world. Corporate scandals – coupled with rapid global communication, a rise in overall education levels, and increased societal expectations of the businesses whose products and services we buy (and of the governments that rule us) – have led to a shift: from a world where people believed what they were told, to one where they asked for information and explanations of what was going on; from there to one where information was not enough, and stronger evidence was required for society to trust business; and thence to a desire in society to be involved with companies, in a true realization of what has become known as stakeholder engagement.

This evolution has occurred, at different speeds and along slightly different paths, in the developed and developing worlds alike. A parallel, related development has been a significant increase in the transparency with which companies operate, not because they were necessarily operating improperly in the past but because society has become less inclined to believe in their propriety without evidence and direct experience of it. A welcome development from this increase in transparency has been an upturn in society's trust and people's preparedness to do business – directly or indirectly – with companies that meet their expectations.

From scenarios to decision-making

Companies have a role to play in re-establishing trust, but they are not the only players. Scenarios such as *Global Scenarios to 2025* are useful, not because they provide a set of futures from which to choose "the most likely" and then prepare for it, but because they establish the right focus for the "what if?" questions – what if other players do this, what if society changes in that way – that are at the heart of any business strategy and planning.

For example: "What if the strategy we propose has to play out in a Low Trust Globalization world? Then we need to be prepared to operate in a world of intrusive checks and controls." Or: "What if our operating environment becomes more like an Open Doors world? Then reputation has to be a primary concern for top management, such that everyone in the operation knows how to act to preserve and enhance it."

The scenarios provide a framework in which to look at the key questions that are going to shape decisions – by policymakers, by other companies, and by regulators. The aim is not to predict what will happen, but rather to enable the user to say: "If this happens, what sort of policy can we expect to be put in place; what sort of strategy is likely to be adopted; will we see cooperation, or will it be everyone for him or herself?"

Even in areas where a company has been operating for some time, it can be difficult for its representatives to see the full picture – and still harder to understand it – in order to make good decisions about how to move forward. We are all limited by personal and group culture and values; by our experiences; by our ignorance (whether conscious or unconscious, voluntary or involuntary); by our expectations; and by human nature's in-built unwillingness to change. It is all too easy to see the world as we would like to see it: benefiting our own interests, not as it actually is.

Decision-making on a technical level is relatively straightforward. There are plenty of useful techniques and devices for understanding and managing the physical realities of a typical energy business. Problems and challenges that are quantifiable can find solutions that are implementable. With the common language of science and technology, those responsible for decisions have a rational basis on which to make such decisions. The emotional content in these decisions is relatively low, and satisfaction levels – "job done" – can be very high.

However, technical decisions must be implemented in the real world. Oil companies do not – and should not – control the social and political environment in which they operate. Among the many other players involved, some may be hostile. Alignment of the interests of oil companies and other players rarely happens automatically. And, however unquantifiable emotions and prejudices may be, it is undoubtedly true that they count as much as the technical realities of operation.

In the real world, there are many players calling the shots (Figure 9.4); and the influence of the oil company, while relatively strong in technical and commercial areas, diminishes rapidly as matters move into the arena of national and international economics and politics.

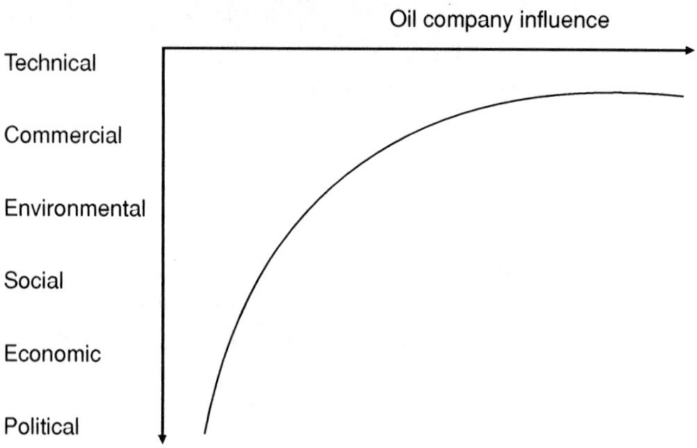

Figure 9.4 In the Real World Other Players Call Many of the Shots

Making the wrong decision, choosing the wrong options, can have serious consequences – not only for the oil company making the decision but for the other players as well. Poor technical or commercial decisions by a company can mean loss of money, reputation, business or even lives. Bonuses may have to be forgone; in the worst cases, bankruptcy can result. When it comes to decisions that affect environmental, social or political matters, where the company has considerably less influence over the final outcome of the decision, the results of poor choices can include serious losses and disruption over decades due to adverse environmental impacts, social change and upheaval, distortion of the economy, and internal tensions and rivalries at any level from the local community to national government.

Moving outside the comfort zone

Conversely, the long-term success of any oil company venture depends on the company's operation and management of its "discomfort zone."

Crucially important are the acquisition and deployment of the skills needed to understand and manage issues and stakeholders.

Capturing and maintaining a "license to operate" demands knowledge of the host country and society – the immediate situation and the direction in which it is moving; of the underlying issues and long-term trends; and of external players – government and multinationals, NGOs (their traditions, interests and networks), and the local and international media. Companies need to understand what all these players expect the business to deliver, as well as the expectations of their own shareholders, managers and employees. In essence: having a license to operate means the government *grants* permission while the community *gives* permission.

Investment decisions, to be effective, should take into account the complex world in which they are being made. In addition to the in-house competences required to assess technical feasibility and commercial/economic viability, companies need to be able to weigh up the external impact of any potential investment: the impact on the environment, on local and national communities, on host government economics, and even on the political regime.

And, at the end of the day, decisions must be implemented in the real world. Here again, both oil companies and other players need to recognize that control of social and political areas, the host government's economic priorities, the interests and short-term focus of other players are all outside the control of any oil company. The consequences of poor decisions are serious for everyone; so mutual understanding and working together, rather than dogmatism and conflict, are in everyone's interests.

10
Oil and Conflict: Lundin Petroleum's Experience in Sudan[1]

Christine Batruch

Lundin Petroleum obtained the rights to explore for and produce oil and gas in concession Block 5A, Unity State, Sudan, in February 1997; it sold these rights in June 2003. During the period in which the company was active in Sudan, it operated in the belief that oil could benefit the economic development of the area and the country as a whole, and that this would have a catalyzing effect on the peace process. The problems that it encountered in the area, however, led the company to constantly reassess its activities, role and responsibilities there. This chapter examines the reasons why Lundin decided to operate in Sudan, the challenges it faced in the course of its activities, the steps it adopted to satisfy both its commercial objectives and ethical concerns, and its efforts to promote a peaceful resolution of the conflict.

Introduction

Lundin Petroleum[2] obtained the rights to explore for and produce oil and gas in concession Block 5A, Unity State, Sudan, in February 1997; it sold these rights in June 2003.

During the period in which the company was active in Sudan, it operated in the belief that oil could benefit the economic development of the area and the country as a whole, and that this would have a catalyzing effect on the peace process. The problems that it encountered in the area, however, led the company to constantly reassess its activities, role and responsibilities there.

This chapter examines the reasons why Lundin decided to operate in Sudan, the challenges it faced in the course of its activities, the steps it adopted to satisfy both its commercial objectives and ethical concerns, and its efforts to promote a peaceful resolution of the conflict.

Sudan's war

Sudan has been embroiled in a civil war that began shortly after it gained independence from the United Kingdom in 1956. It is one of the longest and most tragic wars of modern history: fighting has taken place for nearly 50 years, with a single reprieve between 1972 and 1983.[3] The Government of Sudan and the Sudan People's Liberation Movement/Army (SPLM/A), led by rebel leader John Garang, are the main protagonists in the conflict, which resumed in 1983, although armed militias in different parts of the country have also been involved at various times. The fighting has taken place chiefly in the southern-most parts of the country although other areas, such as the Nuba Mountains region, Unity State and more recently the Darfur region of western Sudan, have also witnessed periods of intense combat.

It is difficult to ascertain the root causes of the war and the contributing factors over such a long period of time. Nonetheless, certain elements have, at various times, played a role in the conflict. They include: (1) the country's extreme poverty – Sudan is ranked among the poorest nations of the world;[4] (2) the religious/racial divide – northern Sudan is mainly Arab and Muslim, while southern Sudan is African and Animist or Christian; (3) the competition for power – political opponents seek a greater participation in power, while regions seek greater autonomy from the central government; and (4) the competition for resources – southern regions contest the government's control over national resources such as water and oil, which originate in the south.

When peace is achieved, it will be easier to determine which of these elements played the decisive role in the conflict and its eventual resolution. What is clear, however, is that the war began years before the presence of oil was even suspected, and it was only after oil was produced that a material basis for a sustainable peace was seen to have been achieved. It is only then that an active, internationally mediated peace process began.[5]

Until that time, Sudan's war had been largely ignored, except from a humanitarian perspective. The conflict was seen as another typical African war: over local issues and involving local parties. The situation seemed insoluble because of the many problems to be resolved and the slight foundations for sustainable peace. However, in the course of the 1990s a number of developments brought the world's attention to Sudan.

The early 1990s had seen the rise of Islamic fundamentalism, which figured prominently in the Sudanese government; the harboring of renowned terrorists such as "Carlos the Jackal" and later Osama bin Laden; and the suspicion that Sudan was linked to the 1995 assassination attempt on Egyptian President Hosni Mubarak. At this stage, Sudan was considered a "rogue nation," which had to be isolated from the community of nations.[6] In the latter part of the 1990s, however, the government adopted certain progressive measures, which the international community interpreted as signals of impending reform and of Sudan's interest in shedding its pariah status. The steps taken by Sudan included the handover of Carlos to French authorities, the expulsion of Osama bin Laden, the purging of key Islamic fundamentalists from the government, allowing the return of political opponents from abroad, the signing of the 1997 Khartoum Peace Agreement with southern opposition groups (see below), improved relations with neighboring countries, and the adoption of a new Constitution and Bill of Rights.

Whereas the United States was reluctant to recognize these efforts immediately, the European Union (EU) decided to engage in a constructive dialog with the Sudanese government because it believed that this approach was more likely to bring results than keeping Sudan isolated. Thus, when Lundin acquired the rights to explore for and produce oil and gas in Block 5A, world opinion regarding Sudan was beginning to change.

Lundin in Sudan

The company's primary concern when considering a new area for activities is geological. If an area presents the required geological profile – that is, if it is assumed to contain oil reserves – Lundin proceeds to study the technical and commercial feasibility of exploiting the oil. In the case of Sudan, the main risk identified in the course of the company's risk analysis was financial. The company decided, however, that the estimated potential oil reserves were important enough to justify the significant investments required for the venture, in particular investments in infrastructure development. It did not identify any legal risks – there were no international or EU sanctions against Sudan that prohibited a European company from doing business there – or political risks – there were no SPLA forces in the concession area, as the civil war was proceeding further south.

The company therefore engaged in negotiations to obtain a license to explore for and produce oil and gas in Block 5A. As in most countries, mining rights in Sudan belong to the central state. Negotiations were therefore held with representatives of the Sudanese Ministry of Energy and Mining (MEM). The terms of the agreement were standard for the trade, with an initial period for oil exploration – in exchange for a work commitment and the carrying of costs – followed by a period of oil production, with cost recovery sought after initial production. The only terms that were specific to the exploration and production-sharing agreement (EPSA) concerned the "Sudanization" of the operations. At the request of the MEM, the company committed itself to hire and train Sudanese with a view to their constituting 50% of the staff within five years of the commencement of operations and 80% within ten years. There was also a provision that the company would carry the costs of its Sudanese partner, Sudapet, which had a 5% interest in the venture.

On its first visit to the concession area, Lundin met with key representatives of the local community, who welcomed oil activities as the only way to promote long-term economic development in their area.[7] They also committed themselves to providing a safe environment for the company to operate in. This commitment arose out of the terms of the Khartoum Peace Agreement, which they had signed with the Sudanese government and which set out the parties' respective rights and responsibilities in the area.[8]

Security, however, proved to be elusive. The prevalence of arms, coupled with the division of tribes into various factions, contributed to making the situation volatile.[9] Within a few years, instances of fighting started to increase. While the company was not directly affected by the fighting at the time, it was nevertheless worried about the safety of its staff and its operations. It was also concerned because of the criticisms that were being directed against an oil consortium situated in a nearby concession. To better understand these developments, Lundin decided in 1999 to commission a socio-political assessment of the area.

The study, conducted both at the Lundin head office in Geneva and in Sudan, was based on an analysis of reports on the political and human rights situation in Sudan, on interviews with company representatives in the head office and in Sudan, and on meetings with members of the government of Sudan and humanitarian organizations. It also included a visit to the concession area.

The report's conclusion was that, despite the lack of evidence of a direct link between the sporadic fighting that had taken place in the concession area and company activities, there was a potential risk of deterioration if the local communities ceased to perceive the role of oil companies as beneficial. The report also noted that in view of the limited positive benefits of the oil activities at the time – revenues were not expected for a number of years, since activities were at the exploration stage – there was a distinct possibility that the local communities would grow disgruntled.[10] The report's main recommendations were that the company should continue to monitor socio-political developments in the concession area and reinforce its existing relationship with the local community.

Community relations

From the time it started its activities in Block 5A, Lundin adopted a proactive approach to community relations. The company not only met with representatives of the local community but also sought to show goodwill towards the population by hiring local staff and improving the infrastructure in the area.[11] The company believed that, if the local population obtained tangible benefits from oil activities, they would be even more supportive of these activities. However, given the lack of required skills locally, the number of people who were hired was minimal and the impact of this effort was limited. Similarly, while infrastructure developments such as bridge and road building increased local mobility, because they had been carried out for operational purposes the company did not consider them as community projects.

The company therefore sought ways to make a more direct contribution to the local community. It initiated a number of projects, which later became an integral part of the company's Community Development and Humanitarian Assistance Program (CDHAP). The projects had three main objectives: (1) to promote better health, hygiene, education and general quality of life for the current and future inhabitants of the concession area of Block 5A, Unity State; (2) to contribute to the economic and social development of the area; and (3) to reinforce relationships between the local community and the company.

Through this program, the company also wished to demonstrate to the local and central authorities that it was concerned with the interests and welfare of the population and was prepared to make significant contributions, despite the fact that it would not obtain any revenues from its activities for a number of years.

In order to ensure that its projects were relevant, Lundin had consulted with a number of local actors, in particular non-governmental organizations (NGOs) that were active in the area. With their assistance, it identified areas of need where it felt it could make a contribution, such as the supply of fresh water, health, education and capacity building.

In the three years Lundin ran CDHAP, it spent over $1.7 million on its various projects. These ranged from the delivery of fresh water by trucks, to the drilling of water wells and the construction of a water filtration unit. In the field of education, Lundin started by supplying educational materials to existing schools and orphanages, then built schools with local materials, and eventually constructed a permanent building to accommodate several hundred children. Through a team of five Sudanese doctors, assisted by local nurses, Lundin provided medical assistance in mobile tent clinics, temporary straw clinics and eventually in a fully equipped permanent clinic that it had built. Similarly, it relied on two veterinarians and local para-veterinarians whom it had trained to tend to local cattle in a vet station and in mobile vet clinics. The capacity-building projects included the creation of a mobile brick factory, a women's development center and a nursery as well as a program for training local people as midwives, para-veterinarians, nurses, brick layers, vector control specialists, computer analysts, and so on. In times of emergency brought about by climatic or security conditions, the company provided ad hoc humanitarian assistance by supplying people with water containers, soap, blankets, mosquito nets and medical services.[12]

From its inception, CDHAP was a constant element of the company's presence in Unity State. Not only were CDHAP staff members often the first to go to projected areas of activities and the last to be pulled out when the security situation deteriorated, but they stayed there even when operations were suspended. During the company's temporary suspensions of activities in 2001 and 2002, services to the community continued to be rendered in the two main towns of the area, Rubkona and Bentiu, and in surrounding villages. Maintaining its presence in the area through CDHAP was the company's way of demonstrating its long-term commitment to the local community and the area.[13]

If CDHAP was the company's most tangible way of showing its concern for the people in the area, it was by no means the only way. Outbreaks of fighting, coupled with allegations that these conflicts were related to oil, led Lundin to reassess its role and responsibilities and seek ways of exercising a positive influence on the protagonists in the conflict.

Internal review

In the latter part of 1999, civil rights activists started to question the role of the Greater Nile Petroleum Operating Company (GNPOC) oil consortium in the conflict.[14] This consortium, which was operating in a concession area adjacent to Lundin's, had participated in the construction of a pipeline linking the southern oil fields of Unity State to the northern city of Port Sudan and was beginning to produce oil. Activists claimed that human rights violations, such as population displacement, had taken place in order to pave the way for the consortium's activities. The consortium consistently refuted these claims. The activists also believed that the revenues obtained by the Sudanese government from GNPOC operations would be used to build up its military arsenal and quash the rebel SPLA.[15]

There was a marked discrepancy between Lundin's first-hand experience in its concession area and reports about what was being alleged to have taken place in the neighboring GNPOC concession. The report commissioned by Lundin confirmed that many elements distinguished the two operations. First, the GNPOC concession area was sparsely inhabited, which gave credence to the claim that population displacement had taken place prior to the commencement of operations, even though this was disproved by satellite images.[16] Second, the local community there was partly of Dinka origin, the main tribal group behind the SPLA; it was therefore conceivable that there could be clashes between them and government forces. Finally, GNPOC operations had started generating revenue for the government, of which little, if any, appeared to be reinvested by the government in the area.

Despite these differences, Lundin recognized that negative perceptions of the effects of oil operations could also come to be applied to its area and therefore decided to set out, in a Code of Conduct, the conditions under which it was prepared to operate.

The Lundin Code of Conduct

The process of development of the Code of Conduct was important for Lundin, as it required the company to assess the role of its business from a different perspective. Lundin's management had always seen (and continues to see) itself as making a positive contribution to economic growth by providing a necessary source of energy. It had also witnessed how oil revenues in undeveloped areas acted as a catalyst for economic development, paving the way for other businesses and international loans.[17] Lundin was aware of the potential negative impacts of its operations on the environment, and took mitigating measures to address them. The socio-political dimension of its activities, however, was not something the company had had particular reason to consider before the Sudan experience. It believed that these were issues beyond its field of competence.

When faced with the possibility that its activities could have a negative impact on the conflict in Sudan, senior management re-examined the company's role from this wider perspective. Lundin established its objective to play a positive role not only directly, in the economic field, but indirectly in the socio-political field as well. As stated by its chairman, the company's aim is "not only to find oil and gas, we are also committed to developing this valuable resource in the best socio-economic manner possible for the benefit of all our partners, including the host country and local communities."[18]

The Code of Conduct was developed after the company had consulted documents in the field of corporate responsibility[19] and after discussions with members of the Board of Directors as well as senior corporate and country-based management. The Code was adopted as a consensus document, which served as a guide for the company's activities worldwide.

The Code set out the company's values, responsibilities and the principles by which it was guided. The company recognized that it had specific responsibilities towards its shareholders, employees, host countries and local communities, as well as to the environment. It committed itself to act in a fair and honest way, to observe both national and international laws, and "to act in accordance with generally accepted principles on the protection of human rights and the environment."[20] After the Code of Conduct had been adopted by the Board, Lundin disseminated it to its employees in Geneva and in Sudan, and to the company's affiliates. It became an integral part of the company's contracts of employment.

The adoption of the Code was followed by other initiatives, such as the publication of the company's policies on health and safety, the environment and community relations. The company also arranged for an awareness session on human rights and developed a human rights primer, explaining the origins of and guiding principles for the protection of human rights and how they relate to business. The company's security liaison personnel in Sudan were provided with information regarding human rights and security, to sensitize them to such issues in conflict situations, and were encouraged to report any violations they witnessed.[21]

The internal dissemination of the Code of Conduct was necessary in order to ensure that the staff understood what the company stood for and what was expected of each and every one of them. It also became the basis for discussions with stakeholder groups in Sudan.

Stakeholder engagement

In the course of developing its Code of Conduct, the company defined more precisely who its stakeholders were in relation to its activities in Sudan. In the first few years of its operations in Sudan, it had cultivated friendly relations with business partners, government representatives at the central and local levels, and community representatives. It also had informal relations with other oil companies and NGOs active in the area. However, it decided that in view of the competing claims being made about the impact of oil in the region, it needed not only to widen the scope of these contacts but also to alter the content of its discussions to include socio-political issues.

The company's early consultations with central and local authorities had revealed a shared view that oil represented a momentous opportunity for the development of the country and the area. Even the humanitarian and development organizations it had consulted at the time recognized this potential, but they remained more reserved as to whether the wealth produced would be properly shared among the population.

This general consensus began to erode, however, when representatives of the local communities whom Lundin had met at the outset accused the Sudanese government of reneging on its commitments under the Khartoum Peace Agreement and decided to resign from their governmental posts. Their decision, coupled with the defection of a local tribal faction to the SPLA, represented a turning point both in the conflict and for the company. Inter-factional fighting escalated into a conflict, which pitted against each other militias that were backed by the two contenders in the civil war – the government of Sudan and the SPLA.

Judging the situation as representing an undue risk to the safety of its staff – the SPLA having then indicated that it considered oil operations and staff as legitimate military targets – Lundin decided to temporarily suspend its operations. It made its resumption of activities conditional upon a peaceful environment, noting that this could only be achieved with the support of the local community.

Lundin also decided to enhance its knowledge of the situation by consulting not only those with whom it had formal relations, such as its partners in the consortium and the government, but also those with particular knowledge of, or interest in, the conflict in Sudan. The purpose of these discussions was to share information and opinions about the conflict and to establish what was required for company operations to resume.

The institutions with which the company met included the following:

The Sudanese Government (host government) and the Government of Unity State (local government)

Discussions with the Sudanese and local governments focused on the means to render the area conducive to oil operations. The company expressed its view that the long-term security required for sustainable oil activities could only be achieved with the support of the local community. Lundin made it clear that, in its view, military action – except for defensive purposes – was not an acceptable option.

The Nuer opposition (local community)

In its discussions with representatives of the Nuer opposition, the company attempted to convey its view that oil presented the best opportunity to achieve sustainable peace and growth in the area and encouraged them to seek a peaceful way to assert their rights to the area.

The Swedish Ministry for Foreign Affairs (home government)

It was important for Lundin, as a Swedish company, to share with the Swedish Ministry for Foreign Affairs (MFA) its views about the situation in Sudan and its approach there. Given the allegations about wrongdoings committed in its area of operations, the company kept the MFA informed of its first-hand experience in the area and the steps taken to address local needs and concerns. As a member of the European Union, Sweden had adopted a policy of constructive engagement in Sudan: the activities of the company fell within this approach, insofar as it ensured that its activities were not affecting the conflict negatively.

United Nations relief organizations (the humanitarian community)

UN organizations were present in Sudan mainly to deliver humanitarian assistance under the umbrella organization Operation Lifeline Sudan (OLS).[22] Set up both in Khartoum to service government-controlled areas and in Lokichoggio, Kenya, to service parts of the country under SPLA control, the OLS had witnessed the unbearable toll of the war on civilians. Its main concern was to have full access to all areas of the country in order to be able to provide humanitarian relief in the case of crises. As the company had itself offered assistance to internally displaced people fleeing from areas of natural or man-made catastrophe, it shared the view of the OLS that unrestricted humanitarian access was required and raised this issue in its meetings with government and Nuer representatives.

The United Nations Commission on Human Rights

The UN Commission on Human Rights had two representatives for Sudan: an in-country representative, whose role was to promote respect for human rights by the Sudanese government and in government-controlled areas; and a Special Rapporteur on the Situation of Human Rights in Sudan, whose role was to assess and report on the human rights situation throughout the country. In 1999 the Special Rapporteur claimed that oil activities had exacerbated the conflict, although he had not visited the oil fields or even consulted with the oil companies. Lundin therefore contacted him to inform him of its first-hand experience and knowledge of the situation in the area and invited him to visit the oil fields instead of relying on secondary, sometimes biased, sources. The eventual visit of the Special Rapporteur to the area took place at such a time and was of such short duration (a mere three hours) that he could not conduct an in-depth inspection. In the course of discussions with company representatives, however, he admitted that the civil war was the cause of the human rights problems and that oil, if properly channelled, could contribute to a sustainable peace.

Non-governmental organizations

The NGOs with a focus on Sudan may be categorized in two broad groups: (1) those that have a permanent presence in Sudan, and assist the population through local humanitarian or development projects; and (2) those that are based outside Sudan, and promote special interests such as human rights, religious rights, development rights, and so on. Lundin was in contact with both groups to exchange views about the situation in Sudan and means to improve it. Not surprisingly, it found that organizations with a humanitarian focus were generally supportive of the company's efforts to contribute to the local communities in its area of operations. They were prepared to talk to company representatives and even work with them on certain projects. When the stigma surrounding oil activities became significant, most chose not to be publicly associated with the company and therefore only a few cooperative ventures continued, on a confidential basis.

Lundin's experience with special-interest NGOs was more difficult. In many cases, views about the situation in Sudan were so very different that discussions rarely went beyond each side trying to convince the other of the correctness of its views. This was particularly true with respect to religious-based organizations, which characterized the conflict as an attempt by Muslims to eradicate the Christian population in the south of Sudan in order to gain access to the oil there. Although the company responded to their claims, in discussions and in writing, it felt that not much would be gained from this effort. These NGOs believed that the cessation of oil activities was a means to achieve peace, while the company believed that oil activities would be the basis for peace.[23]

There were two notable exceptions in Lundin's relations with special-interest groups: Amnesty International, particularly the Swedish branch; and the Church of Sweden. Both organizations believed in the benefits of constructive engagement with companies operating in Sudan and met with Lundin on a number of occasions. Lundin invited their representatives to visit its concession area, but because of its suspension of oil activities and later sale of the asset the visits never materialized. Nonetheless, some of these groups' views and recommendations were taken into consideration and, where appropriate, were integrated into Lundin's business conduct.[24]

Think tanks

The think tanks that had been following and reporting on Sudan for a number of years also considered how oil could act as an incentive for peace in Sudan.[25] Above and beyond the obvious positive benefits of oil for the overall economic performance of the country, they were interested in ascertaining whether oil could be used as a peace incentive. Discussions with representatives of think tanks were dedicated to a review of oil exploration, production and revenue distribution schemes. It was generally accepted that a fair distribution of oil resources was a necessary condition for peace, and in this regard the company drew their attention to the equitable sharing scheme laid out in the Khartoum Peace Agreement.

The media (representing public interest)

When allegations of a possible connection between the war and Lundin's operations surfaced in the press, the company decided that the best way to respond was to invite both Swedish and international journalists to visit its concession area. Until that time, journalists who had reported from the field had been able to do so only with the support of rebel forces; their reports therefore presented only one side of the story. The company believed that if they had the opportunity to visit the area without support or interference from either rebel or government forces, they would have a more balanced and realistic view of the situation. A number of journalists took up the company's suggestion and visited the area in 2001 and 2002. They produced articles for the press as well as video recordings that were aired on both Swedish and Swiss television.

Peace negotiators

As a principle, Lundin refrains from getting involved in the political affairs of a country; it believes that it cannot make a meaningful contribution in this sphere and prefers to restrict itself to its commercial mission. The situation it encountered in Sudan, however, was exceptional, and the company needed to make clear to the protagonists in the conflict that it saw peace as the best means to ensure sustainable oil operations. In this endeavor it relied on the skills and competence of Carl Bildt, a member of Lundin Petroleum's Board of Directors, whose experience as the UN Secretary General's Special Envoy for the Balkans in 1999–2001 was particularly relevant. In a series of trips to Brussels, Cairo, Khartoum, Nairobi and Washington, Bildt met with high-level representatives of the Sudanese government, including the President, his peace advisor, the Minister of Energy and Mining, the Minister of Foreign Affairs, and the main representative of the Nuer community (later deputy chairman of the SPLA), as well as with representatives of the key nations acting as peace mediators, such as Kenya, Norway, the UK and the USA. Bildt delivered the same message to all: oil represented an incentive for peace insofar as oil activities could not be pursued in a war context. He also underlined how oil provided the material basis for a sustainable peace. The company's repeated suspensions of activities were a proof that oil activities could not flourish in a conflict situation, and experience in various other countries demonstrated that a conflict of this nature could not be resolved militarily. In Bildt's view, the parties had to determine for themselves their minimum, not maximum, requirements for the achievement of peace. The mediators' role was to help the parties achieve this compromise by offering them support, in the form of international monitoring and monetary assistance for purposes of reconstruction.

Conclusions: lessons learned

During the seven years in which it acted as operator of Block 5A in southern Sudan, Lundin was faced with a constantly changing environment. The company learned that, despite its desire to restrict itself to a commercial role, it could not ignore either the socio-political developments in its area of operations or the claims – even if unfounded – of a possible connection between its activities and the conflict.

A reaffirmation of its values in a Code of Conduct, a greater involvement in community life, stakeholder engagement and the suspension of activities were the tools adopted by the company in response to the challenges it faced.

In the spring of 2003, the company sold its interest in Block 5A at a profit. The transaction was satisfying not only from a commercial perspective but also from the perspective of corporate responsibility. At the time the company left, active peace negotiations were under way and its community development program was maintained by its successor. This reinforced Lundin's belief that it is possible for business to pursue commercial objectives while meeting ethical concerns, even in areas of conflict.

Notes

1. This chapter originally appeared in A.J.K. Bailes and I. Frommelt (eds), Stockholm International Peace Research Institute (SIPRI), *Business and Security: Public–Private Sector Relationships in a New Security Environment*, Oxford University Press, 2004, pp. 148–60. This chapter is reproduced here with permission from the author and SIPRI.
2. On the independent Swedish oil and gas exploration and production company Lundin Petroleum AB – hereafter referred to as Lundin, or the company – see <www.lundin-petroleum.com/>. Lundin was the operator of Block 5A on behalf of the consortium, which included OMV (Sudan) Exploration GmbH, Petronas Carigali Overseas Sdn Bhd and Sudapet. For a map showing the location of Block 5A, see <www.lundin-petroleum.com/eng/sudan3.shtml>.
3. For an account of recent developments in this conflict, see S. Wiharta and I. Anthony, "Major Armed Conflicts," in *SIPRI Yearbook 2003: Armaments, Disarmament and International Security*, Oxford University Press, 2003, pp. 101–4. At the time of writing, peace negotiations held under the auspices of the Intergovernmental Authority on Development (IGAD) were in their final phase and a comprehensive agreement was expected to be signed by the end of the year. On the peace process, see "Sudan: Peace Talks, Humanitarian Action," <www.irinnews.org/webspecials/sudan/default.asp>; and C.L. Powell, "An Opportunity for Peace in Sudan," October 28, 2003, <www.sudan.net/news/posted/7274.html>.
4. For a discussion of Sudan's economy and the positive impact of oil in the past few years, see the US Department of Energy Internet website at <www.eia.doe.gov/emeu/cabs/sudan.html>.
5. The current phase of peace negotiations originated with the activities of Senator John Danforth, who was appointed by President George W. Bush as Special Envoy for Peace in Sudan on September 6, 2001. See J.C. Danforth, "Report to the President of the United States on the Outlook for Peace in Sudan, April 26, 2002," at <www.sudan.net> (under "Latest news," "Press releases and commentary," posted on May 14, 2002). The oil issue and the means for resolving the conflict are also discussed there.
6. Because of Sudan's perceived connection with international terrorism, the UN and the USA imposed sanctions against Sudan, the former through a travel ban on Sudanese officials and the latter in the form of a ban on the conduct of business in the country by US companies.
7. The company met with Dr. Riek Machar, who, pursuant to the 1997 Khartoum Peace Agreement, was Vice-President of Sudan and President of the South Sudan Coordinating Council (the government representative for the south); with Taban Deng Gai, the Governor of Unity State; and with representatives of the local factions.
8. The text of the Khartoum Peace Agreement, signed in April 1997, is available at <www.sudani.co.za/Documents%20and%20Issues/Khartoum%20Peace%20Agreement.htm>. It was signed between the government of the Sudan,

the South Sudan United Democratic Salvation Front (UDSF) – comprising the South Sudan Independence Movement (SSIM) and the Union of Sudan African Parties (USAP) – the SPLM, the Equatoria Defence Force (EDF), and the South Sudan Independents Group (SSIG).

9. The main tribe in the area is the Nuer tribe, which has five subgroups: the Bul, Lek, Jikany, Jagei and Dok Nuer. In turn, these groups are affiliated with local militia.
10. Oil exploration and production are by nature a long-term activity: it takes a number of years before oil is found, and several more before it is brought into production and sold. It therefore takes years for revenue from oil to accrue to an area, which, in the meantime, has observed construction activity, equipment being brought in, and teams of people going back and forth. In many areas of the world, this poses no particular problem, but in an area like southern Sudan, where the majority of the population live in very precarious conditions, this issue requires special attention.
11. The uniqueness of Lundin's approach did not go unnoticed. Indeed, in a meeting with representatives of an international NGO, Dr. Riek Machar, who had then defected from the government of Sudan, stated that Lundin was different in that it had consulted with the local people and tried to involve them in its activities.
12. For a review of CDHAP activities in 2001–02, see <www.lundin-petroleum.com/eng/comdev.shtml>.
13. This commitment has been passed on to Petronas Carigali Overseas Sdn Bhd, Lundin's successor in the area, which has decided not only to pursue projects initiated by Lundin but also to expand the activities under CDHAP.
14. The GNPOC was at the time a consortium of Chinese, Malaysian, Canadian and Sudanese companies.
15. The consortium contested these allegations. It provided evidence of population growth in the area and divulged the nature of its discussions with the government regarding the use of its facilities for military purposes.
16. The Canadian company in the consortium hired Kalagate Imagery Bureau, a British company specializing in the analysis of satellite images, to ascertain population patterns in its concession area in the 1980s and 1990s. The conclusions were that there was no evidence of appreciable population migration from the area.
17. It had felt this way about Sudan, and in many ways it turned out to be right. Over a period of five years Sudan shed its pariah nation status and became an attractive place for the international business community (sanctioned by the International Monetary Fund).
18. Code of Conduct, "Message from the Chairman." The text of the Lundin Code of Conduct and related documents are available at <www.lundin-petroleum.com/Documents/ot_lupe-code_e.pdf>.
19. These include the Caux Principles, the Global Sullivan Principles, the UN Declaration of Human Rights, the International Labour Organization's Tripartite Declaration of Principle concerning Multinational Enterprises and Social Policy, the Organization for Economic Cooperation and Development's Guidelines for Multinational Enterprises, Amnesty International's Human Rights Code for Companies, the Prince of Wales Business Forum on Operating in Conflict Zones, and so on.

20. Code of Conduct (see Note 18 above).
21. The relevant personnel received information about the Voluntary Principles on Security and Human Rights for the Extractive Sector, available at <www.amnesty.org.uk/business/newslet/spring01/principles.shtml>; Amnesty International's 10 Basic Human Rights Standards for Law Enforcement Officials, <http://web.amnesty.org/aidoc/aidoc_pdf.nsf/Index/POL300041998ENGLISH/$ File/POL3000498.pdf>; the 1990 UN Basic Principles on the Use of Force and Firearms by Law Enforcement Officials, <www.unhchr.ch/html/menu3/b/h_comp43.htm>; and the 1979 UN Code of Conduct for Law Enforcement Officials, <www.unhchr.ch/html/menu3/b/h_comp42.htm>.
22. At that time, the OLS was comprised of 42 intergovernmental and non-governmental development and humanitarian organizations, among which were the UN Children's Fund (UNICEF), the UN Office for the Coordination of Humanitarian Affairs (OCHA) and the UN Development Programme (UNDP).
23. In March 2001 Lundin posted a web report "Lundin in Sudan," which described company activities to date and responded to allegations regarding the nature of the conflict in its area of operations.
24. Amnesty International (AI) had issued recommendations for oil companies operating in Sudan; these were circulated among relevant company staff, as were copies of the 10 Basic Human Rights Standards for Law Enforcement Officials (see Note 21 above).
25. Two US-based think tanks devoted particular attention to this issue: the Center for Strategic International Studies (CSIS), Washington, D.C.; and the Carter Center Peace Program, Atlanta, Georgia.

Part Three
Downstream

11
Natural Gas Vehicles in China: An Answer to High Oil Demand?

David Demers

World oil supply and demand balances continue to exhibit significant strain, with associated upward pressure on oil prices and international competition for available resources. How much global oil supply can be increased and at what price are questions that continue to dominate debates, even as fast-growing economies such as China and India are rapidly expanding their transportation systems and oil-using infrastructure. The rise of "megacities," particularly in Asia, is creating unrelenting demand for private cars, public transit services and comprehensive trucking distribution networks. Of course this increased urbanization has led to worrisome trends in urban air pollution and fears of climate change.

This chapter discusses Westport's experience in China, looking at the interplay between oil price and availability, environmental damage, and the growth of the transportation sector in Asia. The company's view is that global oil demand growth is being driven by transportation growth in Asia, but in the face of apparent supply constraints and risks, both the West, and China too, face severe economic disruption over the next decade due to demand and supply imbalances. There is a need to immediately bleed off some of the pressure from the largest demand blocks: the USA and China vehicular use. One approach is fuel rationing (China is facing serious disruption of supply of all fuels) but high prices are also encouraging conservation in the USA (and enabling more expensive hybrid technology, which will have some impact). China and India are also accelerating their plans to use alternative fuels such as natural gas. The very large investment in liquefied natural gas (LNG) facilities is offering China a unique opportunity to launch large-scale transportation sector projects, which could change the oil demand picture considerably over the next decade.

Natural gas will pave the way for more advanced gaseous fuels such as hydrogen, which could further improve the situation. It has already been demonstrated as an excellent vehicle fuel in many parts of the world, while technologies for the use of hydrogen continue to be developed over the long term.

China's energy situation in the world

China's economic growth in the past few decades has been nothing short of breathtaking, with GDP growth averaging over 8% compounded over 25 years. This spectacular performance, sustained over such an extended period, from such a large economy, is unprecedented.

What comes next? It is clear that the government's mission is to continue this performance, to bring further improvements in income and lifestyle for China's citizens. China's planners hope for average annual GDP growth of 7% above the OECD average growth rate. Over 25 years this would substantially narrow the gap between China and OECD economies, and of course China would pass the USA as the world's largest economy, albeit still with lower individual income.

This is an impressive goal. Can it be achieved? China's track record suggests we shouldn't dismiss the possibility, but there are certainly daunting obstacles and challenges to be overcome. Raw material shortages and associated economic inefficiencies abound, and in particular energy shortages are seen throughout the economy, despite prodigious investment in mega-projects such as the Three Gorges Dam and the West–East natural gas pipeline. Economic benefits are not being uniformly felt throughout the country, leading to social tensions. Environmental issues are formidable, and still growing. The number of qualified administrators and business managers available to deal with these challenges is limited.

And, of course, China's economy has become large enough to have a material impact on the rest of the world. China has become one of the world's largest consumers of commodities of all kinds, and in particular has emerged as a major competitor for oil in the world markets. In addition to the internal challenges associated with such rapid economic growth and social change, China is becoming a global superpower, which brings its own competitive tensions and risks.

Global oil demand growth has leapt to 3% per year, up considerably from the historical rate, which has typically been less than 1% over the past 25 years. Approximately half of the global increase can be attributed to Asia, where growth has been primarily driven by the Chinese economy. This has put severe strain on the global oil supply chain and as a result oil prices have tripled over the past few years. Although there are many different scenarios for future oil supply and demand, the only thing that seems certain is that demand will continue to increase, with supply being somewhat uncertain and subject to disruption. This is a recipe for continued price volatility and potentially significant price spikes.

Oil producers are celebrating their good fortune, but consuming economies are shocked by the sudden and sustained price hikes, despite the previously reassuring forecasts from economists. Sustained high oil prices will put a brake on global economic demand unless we can use the opportunity to improve energy efficiency, and also begin to diversify away from petroleum as the world's primary energy source. It is not inconceivable that these economic drivers will spur innovation around the world. But China's situation and prospects are particularly interesting.

Energy security

China is exploiting every possible energy source as part of its economic development. According to the International Energy Agency (IEA), its primary energy demand is projected to grow on average 2.6% annually – more than doubling its 1,242 million tons of oil equivalent (Mtoe) in 2002 to 2,539 Mtoe by 2030. Nuclear and natural gas are the fastest growing segments over the next 25 years, growing at an average of 9% and 5.4% respectively. But oil, at 3.4% growth will be a problem of global scale.[1]

In 2003 China was the world's second largest consumer of petroleum products after the USA, surpassing Japan for the first time; and it continues to grow more quickly than the rest of the world. Its petroleum processing capacity is now second in the world and domestic production continues to grow. However, China's total energy demand of 5.5 million barrels per day (MMbpd) in 2003 is projected to reach 13 MMbpd by 2025. Furthermore, 2003 import requirements of 2.2 MMbpd are expected to rise to 4 MMbpd in 2006, 6 MMbpd by 2020 and 9.5 MMbpd in 2025. By then, the country will have drastically increased its oil import dependence from currently around 40% to then 70% of total demand.

To put this into context, the entire output from the Canadian oil sands may possibly reach 3 MMbpd by 2025, more than tripling current production. This will require investment of over $150 billion to develop this resource. But China, looking for almost 10 MMbpd of new oil capacity, will need all the output of three such global projects to meet its demand over the next 20 years. We simply don't have the oil resources to do this, disregarding the enormous capital investment required.

The cost of projected oil imports is also a severe burden on the rapidly growing Chinese economy. Adding to the severity of this issue is the concentration of the sources for oil in the Middle East. Any disruptions in this supply chain would have extreme adverse consequences on a nation growing as rapidly as China.

Pollution

The environmental challenges are no less significant. Recent industrialization and vehicle growth have resulted in poor air quality in most urban centers. Since 1990 there has been a sharp increase in ambient NO_x levels, coinciding with the increase of vehicles. A report released in 1998 by the World Health Organization noted that of the ten most polluted cities in the world, seven were in China. If current trends continue, by 2010 it is estimated that 64% of pollution in China will come from vehicle exhaust. And the 2010 vehicle population could double again by 2020.

China is already the second largest contributor, after the USA, to global energy-related CO_2 emissions and is expected to account for 26% of new global emissions between 2002 and 2030, exceeding the increase in emissions from all the OECD countries combined. The transport sector will see the fastest growth of emissions in China.

Urbanization

Currently 31% of China's population lives in urban areas, compared to 78% on average for developed nations. By 2010 this figure is projected to rise to 40%, translating to more than 400 million people. Urbanization has created a greater demand for transit and intercity commercial vehicles, resulting in increased urban pollution, particularly nitrogen oxides and particulate matter.

Vehicle growth

China's 30 million vehicles on the roads are projected to more than quadruple over the next three decades, and Chinese vehicle ownership is projected to reach more than 90 per 1,000 persons by 2030. It must be noted that this would still be only a tiny fraction of the US economy's vehicle density, which today has more vehicles licensed than there are registered drivers: about 800 vehicles per 1,000 persons. Energy demand for road transport in China is projected to increase by 6.5% per year from 2002 to 2012. From thereon and up to 2030, the vehicle stock and annual oil demand will more than double. The economic impact will clearly be different for a $40/bbl than a $100/bbl oil price scenario.[2]

China's challenge: transportation energy

The primary driver of energy growth in China, and hence oil consumption, is the rapid and impressive commitment to growth of on-road transportation. Transportation is already by far the largest consumer of oil, and the only sector that is growing its oil demand. As cars and trucks proliferate, gasoline and diesel fuel consumption rise accordingly. Growth in the transportation sector is a necessary prerequisite to the wider economic development plans in Asia. Therefore, without access to energy to fuel the transportation sector, economic growth at the planned pace will not be possible. But slower growth is equally problematic for the Chinese leadership, who face social disruption if economic benefits don't quickly spread to the more rural west of the country. They need to find a way to keep growth going in a more sustainable way.

Yet the cost of oil, coupled with the apparent supply challenges, and the high urban pollution damage caused by conventional vehicles raise major concerns. State leaders have recognized the growing economic burden of pollution and are increasingly aware that economic growth is jeopardized by the environmental situation.

There are two realistic paths forward.

First, priority should be put on reducing fuel consumption and emissions from both new and existing vehicles, particularly trucks and buses.

Second, China needs to prioritize development of alternative fuel vehicles, particularly in high fuel-use applications like trucks and buses. A single heavy-duty truck will consume, on average, more than 80,000 liters of diesel fuel per year, more than 30 times the average consumption of a taxi and double that of a transit bus. As shown in the model discussed below, converting half of the heavy duty truck fleet in China could on its own displace almost 14% of the oil imports in 2020. Heavy duty trucks represent about 9% of the commercial vehicle population in China, so even more displacement is clearly possible with this strategy.

Unfortunately technology change in the automotive industry is slow; the nature of the industry means that incremental change, often driven by incremental regulatory change, is much more common than truly revolutionary, disruptive technologies. Product development costs are too high and market risks too large for most manufacturers to contemplate speculative ventures involving new fuels. Widespread adoption of alternative fuel vehicles will be driven by local economic forces, but requires government policy initiatives to instigate activity in the industry. In the longer term, only economics – meaning total cost of ownership, including original price, fuel and maintenance costs, and durability – can deliver sustainable changes in energy markets.

Technology change in the automotive sector has been driven by two external market factors, both heavily influenced by government policy:

Environmental regulation of tailpipe emissions, by-products of the internal combustion engine, requires manufacturers to certify that their products do not exceed the applicable emissions caps.[3] Currently, there are three major jurisdictions to consider: California, typically the toughest in the world and the leading indicator of future regulation; the US EPA (US states other than those adopting California standards); and the European Union (which is typically adopted in Asia). Each jurisdiction shares similar principles but the details are materially different, requiring products to be configured differently for each market.

Most other countries adopt, or loosely follow, one of these standards. Some emissions, such as greenhouse gases, are not regulated specifically in this way but are under intense scrutiny. Transportation is recognized as one of the largest sources, and the fastest growing source, of greenhouse gas emissions.

Fuel cost and availability. Many countries and territories (India, Russia and Brazil for example) have large natural gas reserves and the desire to export any available oil in order to earn foreign currency. Investment in infrastructure and vehicle product availability has led to enough critical mass that the industry is seeing high growth and penetration. Natural gas typically can be found at 50% of the price of gasoline or diesel fuel in these locations. In Europe and the USA, increasing oil prices are encouraging more fuel-efficient engines such as diesels and gasoline hybrids, but also renewed interest in natural gas. Some countries, including the United States, have recently established significant subsidies for alternative fuels such as ethanol and natural gas in order to reduce dependency on imported oil.

Both factors therefore imply government policy action to determine the desired energy future. Various technology approaches are possible to deliver the benefits.

Potential for efficiency improvements

Substantial energy efficiency improvements could be achieved through widespread substitution of advanced diesel engine technology for vehicles that currently use gasoline, and, for urban light duty applications, diesel hybrids. These are proven technologies, with established performance and economic parameters. Compared to gasoline engines, diesels will typically deliver 25% better fuel economy.

This approach does not come without costs, of course. The technology for advanced electronic direct-injection systems for these engines is not yet widespread in China. And diesel engines bring their own environmental challenges – relatively high nitrogen oxides and particulate matter emissions, compared to gasoline engines. Emissions-control equipment will add additional costs. Tailpipe after-treatment technologies are emerging globally as both US EPA and European emissions regulations require substantial reductions in emissions over the next five years.

What about hybrid-electric systems? After an initial marketing success by Toyota and Honda many automotive manufacturers are adding hybrid technology, particularly in North America. In practice, fuel economy benefits have been much lower than hoped – typically in the 10–15% range. Hybrids offer most benefits in continuous stop-and-go duty cycles, but at higher speeds – highway driving, for example – fuel economy can be worse than that of a conventional vehicle. The additional cost of the system is unlikely to be recovered through fuel savings within the realistic life of the vehicle. Still, there is no doubt a place for hybrid technology in the future vehicle population mix.

Regardless of the fuel or drive technology, we could save considerable amounts of fuel by putting vehicles on a diet. Smaller, lighter vehicles are eminently possible and the simplest approach to reducing fuel consumption. Advanced materials would allow vehicle designers to cut weight by up to 40% while maintaining structural integrity, durability and safety. Weight reduction would have an almost linear relationship with fuel economy improvement.

All three of the suggested approaches – increased use of diesel-cycle engines (albeit with advanced emissions-control technology), increased penetration of hybrid-electric systems in appropriate applications, and use of advanced materials to reduce weight – would involve new technologies for the Chinese automotive industry, and inevitably would increase vehicle costs. Emissions-control equipment and energy efficiency will be expensive. The relative cost of the new technology will need to be weighed against the cost of inaction and the projected value of fuel savings.

However, the Chinese market offers two unique advantages: the domestic economy alone offers the world's largest potential production volume, and the manufacturing base is sophisticated, flexible and already the world cost leader. Therefore, rapid cost reductions could be possible with widespread proliferation of new technology into the Chinese market.

Adopting new technologies would also enhance Chinese companies' global competitiveness and productivity.

Potential for alternative fuels

Various fuels are possible substitutes for diesel and gasoline in vehicles. Let us briefly consider the leading choices, the primary criteria being that adoption should materially reduce oil consumption in China over the next two decades.

A comprehensive analysis of the viability and benefits of the various alternatives was published in 2004 by the South Coast Air Quality Management District. The conclusions regarding environmental benefits are summarized in Figure 11.1.

The three most promising alternative fuels are biofuels (synthetic fuels derived from organic feedstocks), hydrogen and methane (natural gas). Of these, the best environmental performance is achieved by hydrogen fuel cell vehicles powered from renewable sources, but this technology is not yet ready for deployment. Commercially available natural gas engines offer 90% of the benefit of hydrogen compared to current diesel engines. Other fuels, including biodiesel and ethanol blends, offer less benefit than simply controlling emissions with after-treatment technology.

Let us consider the fuels from an economic and oil security perspective.

Biofuels

Many synthetic fuels, typically used as blending additives with conventional gasoline and diesel fuel, are increasingly being produced around the world from various feedstocks. Biodiesel, produced from vegetable oil or animal fat, is commonly seen as "B20" (20% synthetic, 80% petroleum-based diesel); ethanol ("E85") is also common as a gasoline variant. (E85 is 15% gasoline, 85% ethanol.) Landfill methane and coal bed methane are seen as new sources of natural gas. Synthetic natural gas can also be produced from organic scrap such as waste wood.

On the surface, deriving fuels from waste oil or renewable agriculture products is a very attractive idea. As always, the details include some challenges. Processing scale and economics are obstacles, and environmental benefits can be limited – particularly when considering the total energy lifecycle of the fuel production process.

Recovering biodiesel from waste oils is entirely dependent on the nature and quantity of the waste stream being considered. Processing economics mean that local supplies are required; this naturally limits the

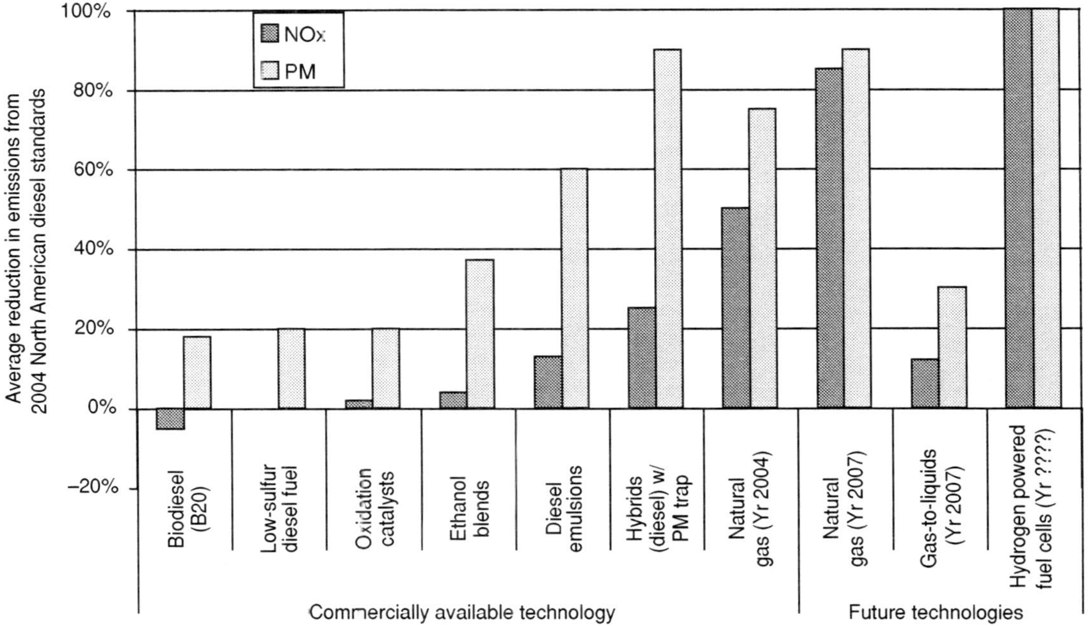

Source: Adapted from South Coast Air Quality Management District

Figure 11.1 Environmental Assessment of Alternative Fuels in 2004

volume of fuel that can be produced. As shown in the analysis above, biodiesels generally increase nitrogen oxide emissions – the primary smog precursor – and because biodiesels can degrade engine materials, some manufacturers limit biodiesel blends to no more than 5%.

E85 is increasingly available, particularly in the USA, although conventional gasoline engines must be modified to operate on this fuel. It is well subsidized in the USA and may develop sustainable critical mass.

Complete energy and emissions lifecycle analysis is needed to assess both the benefits and costs of all synthetic fuels. It would seem that in some projects more energy is used in producing and processing the feedstock than the fuel produced – clearly not a viable long-term proposition. Again, we need to tailor engines to the resulting fuel to get the best results, which adds to the economic and technical obstacles to widespread adoption.

With environmental benefits being neutral at best, and with limited economic benefit, the primary reason to explore biofuels is to reduce dependence on imported oil. This alone is sufficient to ensure that biofuels will play their part in the future fuel mix, although it seems unlikely that they can be a material factor in reducing demand for oil in the near term.

Hydrogen

In the eyes of many environmentalists hydrogen is the logical and necessary future primary energy carrier. Nevertheless there remain many daunting obstacles to its widespread use in the energy economy and in particular as a vehicle fuel.

Hydrogen is widely used in many industrial processes – for example, the production of ammonia and methanol, and the hydrogenation of fats and oils. It is highly flammable, which is why it makes an excellent rocket fuel. It has also been demonstrated as a fuel for more conventional vehicles – for example, in BMW's 7-series cars with hydrogen internal combustion engines and in buses powered by Ballard PEM fuel cells. Westport has also demonstrated hydrogen blended with natural gas in conventional natural gas buses and refuse trucks in California and in Vancouver.

The primary advantage of hydrogen is environmental: if produced from renewable sources, such as the solar-powered electrolysis of water, the total production and vehicle emissions are virtually zero. Unfortunately, very little hydrogen is produced this way today. Total lifecycle emissions, even with a zero emission vehicle, can be higher than using conventional fuels.

Hydrogen is much less dense than conventional fuels, requiring much more space and weight for fuel storage on a vehicle in order to achieve comparable performance and range between refueling. It is also a very reactive element: many conventional vehicle components would be damaged by exposure to hydrogen. Whether powered by hydrogen engines or fuel cells, significant changes to vehicle components, maintenance and operating patterns will be required. In practice, hydrogen vehicles – both those with engines and those with fuel cells – have significant performance, efficiency, and durability challenges that must be resolved before commercial deployment would be viable.

The largest barriers to widespread hydrogen use, though, are cost of fuel and the capital investment required to create safe and reliable hydrogen-capable distribution infrastructure. Realistic scenarios for large-scale renewable hydrogen production suggest a delivered cost of three to five times that of diesel or gasoline, even at current high oil prices. This estimate also presumes no recovery for the initial capital costs of production and distribution equipment.

Although these problems are not insoluble, it will take time, resolve and serious amounts of invested capital to achieve sufficient volumes of hydrogen to displace material amounts of oil. This also presupposes that the many deficiencies in performance, reliability and efficiency in current demonstration vehicles will be successfully resolved within the same time period.

Natural gas (methane)

Natural gas has been used as a vehicle fuel for many years and is now widely established around the world. Although still less than 1% of the global transportation fuel market, this position makes it by far the leading alternative to petroleum fuels.

Natural gas, like hydrogen and biofuels, has advantages and disadvantages compared to conventional fuels, but it has one unique advantage – it is typically significantly cheaper than oil, and much cheaper than other alternatives. It is more broadly available than oil around the world, and we have more of it. Although there is some evidence of a Hubbert's peak for natural gas in the coming decades, we are still finding new large gas resources – unlike oil, where we are using more than twice as much as we are discovering each year.

Natural gas is also significantly cleaner than gasoline and diesel fuel in transportation applications. Because of methane's high hydrogen-to-carbon ratio (CH_4), it produces much less particulate matter and CO_2 than conventional fuels, and its lower combustion temperature ensures it produces less nitrogen oxides. Like ethanol, conventional gasoline engines can be modified to burn natural gas, significantly reducing the engineering and development challenges involved in the adoption of an alternative fuel.

On the other hand, natural gas fuel storage – like hydrogen, although not as extreme – requires more space and weight than oil for equivalent range. Natural gas also requires completely different fuel station storage and dispensing technology. Typical fuel stations use natural gas from the urban distribution network, compressed to 3,600 psi with a large "fast fill" compressor. Pressure tanks on the vehicle accept the high pressure gas from the station. This approach, "compressed natural gas" is abbreviated CNG.

High fuel-use vehicles can adopt liquefied natural gas as an alternative to CNG, which requires only about 1.7 times the space of diesel fuel. This is a realistic compromise for vehicles such as trucks, which have room for external tanks. CNG buses are often configured with roof-mounted CNG tank packages, although this is impractical for trucks or passenger cars. Although passenger cars may require more frequent refueling, there are other potential advantages: In the USA Honda has been offering a home refueling device, which means that natural gas vehicles never need to visit a fuel station.

Relative to conventional fuels, environmental compliance, trade-offs and after-treatment technologies are less costly with natural gas, which makes it the fuel of choice in transportation until the arrival of renewable hydrogen fuel cells.

By partnering with existing engine and vehicle manufacturers and basing new products on an existing vehicle platform, the changes needed to move to natural gas can be minimized: Production, distribution and product support are all simplified and the ability to leverage existing manufacturing capacity maximized.

With appropriate engineering work, natural gas engines offer both environmental benefit and fuel cost saving that is sufficient to offset the increased technology cost of the needed advanced combustion systems. Therefore, we believe that this approach is the most promising solution in order to achieve significant market penetration.

As oil prices and the cost of environmental compliance rise, the incremental cost of new natural gas technologies becomes increasingly attractive. As the volume of natural gas products rise, quality will improve and costs will fall. At some point, economics will drive a rapid shift to the new fuel regardless of the environmental incentives.

These two factors are reinforcing each other to such an extent that in some jurisdictions and in some applications – such as transit buses in North America – Westport believes that natural gas engine products sold through its 50:50 joint venture with Cummins Inc. are already the economic leader compared to diesel fuel. In 2007, as diesel vehicles in North America require complex emissions technology to meet new standards, this trend will accelerate.

In the USA, China and Europe, by 2010–12 – even if oil prices fall – natural gas vehicles may have an unequivocal economic lead in most traditional diesel applications. In that event, much like the US trucking industry's shift from gasoline to diesel fuel in only a few years after the first oil shock, a rapid consumer shift may be possible.

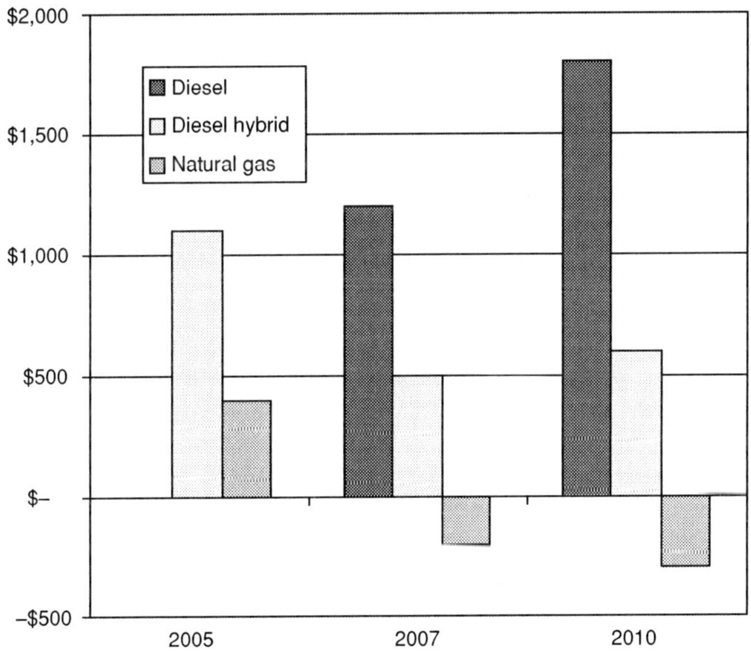

Source: Adapted from Cummins Westport Inc.

Figure 11.2 Total Incremental Lifecycle Cost Modeling – US Urban Bus 2005–10 Compared to 2004 Diesel

In the near term, there seems no better choice for China as a large-scale substitute for imported oil. So let us consider this idea in more detail.

Natural gas in China: an overview

China has large domestic natural gas reserves, has invested billions of dollars in natural gas pipeline infrastructure, is constructing liquid natural gas import terminals all along its east coast and is evaluating the feasibility of international pipelines. It is already experienced with natural gas as a transportation fuel, ranking in the top ten internationally.

Natural gas has typically been allocated to home heating uses, and to power generation. Although China also faces persistent demand–supply imbalances for electricity, a strong nuclear program, the Three Gorges Dam project, continued development of clean coal, and emerging renewable sources offer several viable alternatives that can be implemented with sufficient scale to meet its needs. However, natural gas remains the only realistic alternative to oil-based fuels for transportation.

China's natural gas demand is forecasted to quadruple over the next three decades, leading developing Asia – the region with the highest natural gas growth on a percentage basis worldwide. Its consumption was estimated at 53.3 trillion cubic feet in 2002.[4] The largest reserves include the western Junggar, Tarim, Turpan Harmi and Quaidam basins as well as the central and north-central Sichuan, Ordos, Bohai Bay and Songliao basins. But their location requires significant investments in pipeline infrastructure to carry gas to eastern cities. The West–East natural gas pipeline (the second largest Chinese infrastructure project after the Three Gorges Dam) is transporting gas from the westernmost regions of the country to Shanghai and cities in between.

Plans also exist for the construction of many liquefied natural gas (LNG) import terminals along the east coast. The US$880 million Guangdong terminal became operational in 2006, and a second terminal in Fujian will be open in 2007–08. LNG allows natural gas to be shipped globally, much like oil, removing pipeline restrictions, and it also allows the purchaser to diversify its energy suppliers to countries outside the Middle East. It has the added advantage of requiring little or no refining capacity at the import site, which has been a major bottleneck in China when dealing with diesel imports.

LNG import capacity has increased dramatically in just the past two years. In 2005, Cambridge Energy Research Associates (CERA) counted 13 proposed and two partially approved LNG import terminals in addition to the four under construction or expected to commence construction soon in Guandong, Fujian, Zhejiang and Shanghai. These plans will take China from being a country with no LNG importing capacity, to the second largest in the world in ten years.

China is also increasing its potential to import natural gas by pipeline. Several pipelines from eastern Russian fields in Sakhalin or Irkutsk are being considered to deliver gas into Shenyang in northeastern China, and thought is being given to linking pipelines from Kazakhstan and Turkmenistan into the West–East pipeline.

Natural gas as a transportation fuel in China

Over the past decade China has encouraged the adoption of natural gas as a fuel for taxis and city buses. As part of its commitment to the Green Olympics, Beijing has more than 2,500 transit buses operating on natural gas. Virtually all of these vehicles store the fuel in compressed natural gas tanks (CNG). LPG has also become well established, particularly in the Guangdong area. These fuels have been successful because they are inexpensive, readily available in many local markets, and relatively clean-burning.

In what segments of the transportation economy are the best candidates for natural gas fueling? After studying this issue for a decade, Westport has concluded that the highest economic and environmental value – and hence the most rapid adoption and the highest fuel displacement – will emerge if we can switch high fuel-use applications such as buses, trucks and off-road vehicles to natural gas. Because of the significant economic benefits, we believe market penetration could happen quickly in these segments, much as the first oil shock in the 1970s inspired the rapid conversion of the trucking industry from gasoline to diesel fuel.

Heavy-duty trucks typically consume 80,000 liters of diesel fuel per year, compared to 40,000 liters for a bus, or about 2,300 liters of gasoline in a taxi. Off-road equipment such as mining trucks can consume 1,500,000 liters of diesel fuel in a year of operation. Obviously if there is an economic benefit, larger fuel-use applications have a significant interest in moving to natural gas fuel. Each vehicle converted is far more productive in terms of displacing oil imports than if the same resources were directed into converting passenger cars. Finally, refueling infrastructure is much more productive and justifiable if it is constructed at the same time as a large fuel-use fleet is coming online.

Challenges to adoption of natural gas as fuel

Unfortunately, changing to a gaseous fuel from conventional gasoline and diesel fuel in an automotive application is a complex systems problem. The particular challenges are:

- Developing the complex combustion controls, fuel handling equipment, and the component supply chain for gaseous fuels like methane (natural gas) or hydrogen. The engines, although produced in low volumes, must still meet all customer requirements for performance, durability and reliability, and efficiency despite burning the new fuel, and be cost-competitive. On-vehicle fuel storage systems, which are much more complex, take up more space, and are much more expensive than for conventional fuels.

- Building and operating convenient and cost-effective local vehicle refueling infrastructure with reliable fuel supply.
- Training and supporting prospective fleet users as they consider the risk of change from a proven and known fuel to a new technology, no matter how attractive.

Without successful and visible, conclusive resolution of all three issues for each specific target market, limited or no sales will occur. Prospects for a particular market and application depend on how successfully government and industry leaders can establish a functioning complete value chain for the customer.

Availability of natural gas vehicles

Most of China's existing natural gas vehicles (NGVs) are light to medium-duty vehicles such as taxis and shuttle vans. The reason for this is that aftermarket conversion of gasoline engines to operate on CNG or LPG is relatively simple, and operators are motivated to change by the lower price for natural gas fuel. Typically regional aftermarket providers have appeared to develop and install conversion kits for local customers. This pattern is as true in China as it is in California. However such conversions are generally poorly engineered, and introduce undesirable performance and fuel economy trade-offs. This approach does not meet the needs of more sophisticated customers, who expect the high quality and support of well-engineered new vehicles from recognized original equipment manufacturers (OEMs).

Typically, OEMs do not see enough incremental margin opportunity to justify the comprehensive product development, certification and marketing costs needed to introduce alternative fuel products. Such products are usually seen as niche opportunities, at least in the near term. There is also the perception that NGVs will simply cannibalize existing product sales, with the overall effect of reducing margins and increasing expenses. OEMs also do not have insight into the development of fueling infrastructure and relative energy costs. As a result, product availability has been limited and slow to develop.

Nevertheless, some Chinese OEMs – typically transit bus and taxi manufacturers – now offer natural gas as an option in their product line.

Chinese component suppliers have emerged to respond to the demand from OEMs and customers for NGV specific components – fuel tanks, controls, and fuel systems. In the medium and heavy-duty segments, natural gas has successfully penetrated transit bus and refuse truck markets around the world; several OEMs offer factory NGVs for these markets. Westport's joint venture with Cummins, Cummins Westport Inc., has the broadest global presence with more than 40 different OEM vehicles offered in 18 countries with CWI's natural gas engines as a factory option.

Technology for heavy duty natural gas engines

A significant technical obstacle in heavy duty applications, where diesel engines predominate, is that natural gas cannot auto-ignite in engines with normal combustion chamber design, unlike diesel fuel. Some form of ignition assistance is required. As a result, virtually all natural gas engine manufacturers have opted to forgo compression-ignition and adopt spark plugs for ignition. Typically, spark ignition engines will have lower efficiency – often 25% lower – than an equivalent diesel engine, and reduced performance (particularly torque). Although more fuel is consumed with this approach, the large price differential between the two fuels more than makes up for the penalty. Nevertheless this is not optimum and the performance limitations of spark ignition engines generally restrict their application to transit buses and smaller vehicles.

Two other approaches have been demonstrated that could be viable in the trucking industry: first, dual fuel diesel engines, where some (up to 80% but typically less) fuel substitution is accomplished by mixing natural gas with intake air, followed by traditional diesel fuel injection; and second, direct injection of natural gas into the combustion chamber in the same fashion as diesel engines, but with ignition assist in the form of diesel fuel pilot or glow-plug. The advantage of direct injection is that efficiency and performance are optimized – matching or even surpassing equivalent diesels; but the disadvantages are relatively high complexity and cost. Nevertheless, with the global increase in diesel prices, economically attractive heavy-duty solutions are now visible.

China's automotive industry

Although virtually all of the leading international automotive companies are active in China, the government-controlled market is showing a strong preference for local brands. This is driven primarily by price competition, the rapid development of industry and the government's preference for supporting the growth of its own automotive industry. Chinese automotive companies are seeking ways to differentiate their products and in particular to reach technology alliances to introduce "green" vehicles.

China has the largest commercial vehicle market in the world, with the highest growth rate. Given the region's rapid economic growth, it is inevitable that China, and the other Asian economies surrounding it, will become the world's leading market for vehicles. With this huge domestic manufacturing and customer base and fiercely competitive pricing, it seems certain that Chinese vehicle manufacturers will capture substantial share in the global market once they reach global quality and regulatory compliance standards.

China's heavy and medium truck sales are the largest in the world, exceeding those of the USA, Europe and Japan. In addition, the Chinese transit bus market of approximately 20,000 units is three to four times that of the USA or Europe and is growing rapidly with increased urbanization. In absolute numbers, China's light truck and passenger car markets are small in comparison to the developed markets, but are experiencing much more rapid growth.

In China the natural gas price advantage over diesel typically ranges from 20% to 50% on an energy basis. The advantage over gasoline is even greater, allowing the vehicle operators to save significantly on operating costs by using natural gas.

Deploying heavy-duty natural gas vehicles in China

Westport has worked with a consortium of local companies to propose a number of "green corridor" initiatives. The objective is to create an economically sustainable concentration of vehicles and associated infrastructure in high-use areas, such as truck traffic from ports and major distribution centers.

One such proposed project would enable liquid to compressed natural gas (LCNG) infrastructure between the two Olympic cities of Qingdao and Beijing in 2006–08. It requires that LCNG stations are constructed to support new technology and locally manufactured trucks and coaches be introduced and demonstrated along this corridor. Over the next five years, additional infrastructure will be needed to accommodate the growth of vehicle fleets.

Aside from the obvious Olympic marketing advantage with this "green transport" program, the Shandong Province is home to one of China's largest ports (Qingdao), has a substantial commitment to green energy,

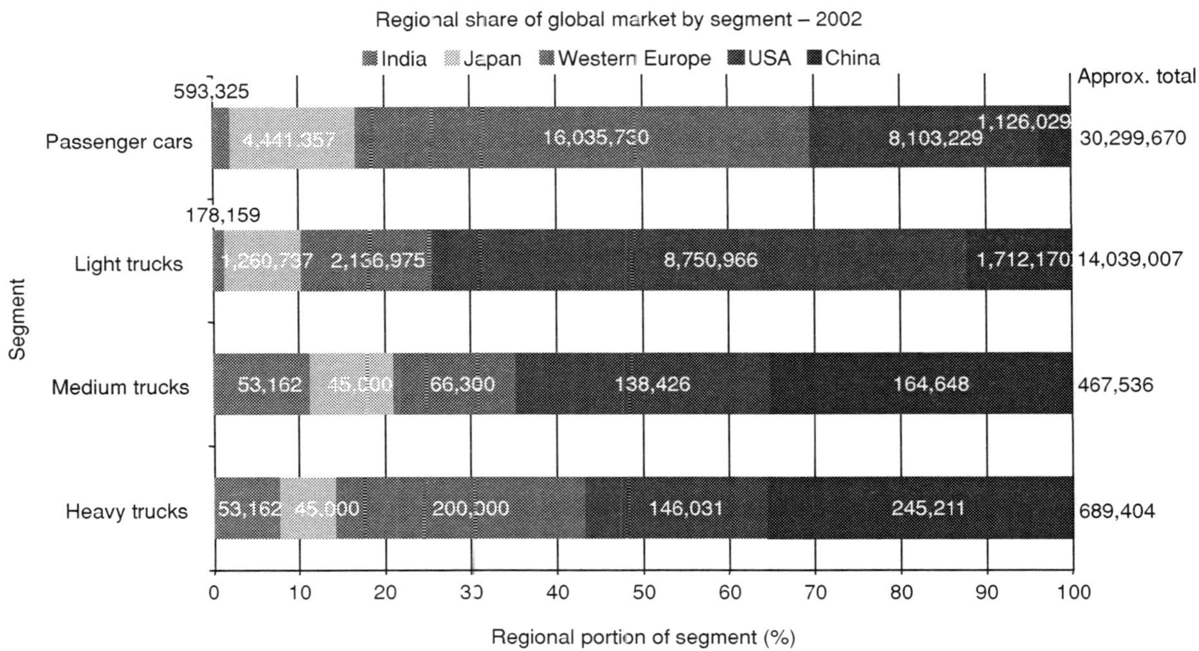

Source: Adapted from World Markets Research Centre & Westport Analysis

Figure 11.3 Automotive Market Relative Global Share (2002)

has current and future access to LNG, and is home to Weichai and Sinotrucks, two of China's largest manufacturers of heavy-duty diesel engines and trucks, respectively.

Westport has recommended standardized use of LCNG technology for refueling infrastructure. LCNG is an emerging solution for rapid and economical development of natural gas transportation infrastructure. Unlike conventional CNG compressor stations, LCNG stations store fuel in liquefied form (LNG) and then can either dispense LNG directly (for heavy duty vehicles) or vaporize the LNG and dispense CNG (light duty vehicles). The major advantages of LCNG infrastructure are:

- Both LNG and CNG vehicles can be supported from the same station.
- Station locations can be independent of the natural gas pipeline, giving much more flexibility in siting and allowing natural gas vehicles to function even where pipeline supply is not yet established. LNG is delivered by truck like conventional diesel or gasoline.
- Multiple LNG supply routes and relationships are possible, allowing flexibility and efficiency in the supply chain.
- On-site fuel storage is available in case of an interruption of pipeline supply.
- Fueling-station design is similar to that of a gasoline or diesel station.
- Operating and maintenance costs are significantly lower, with about 85% less power required than for traditional CNG stations.
- LNG can be offloaded from large-scale, low-cost import terminals and used directly to fuel vehicles, reducing distribution costs.

Westport's Green Corridor projects assume that Chinese partners will emerge to develop the needed infrastructure and to produce low-cost, high-quality natural gas vehicles for domestic customers. Westport has relevant experience in North America, where it has developed an integrated approach to development of this market. A brief review of the primary elements of this strategy may be useful.

North American experience: Cummins Westport

In 2001, Westport and Cummins, Inc. – one of the world's leading diesel engine manufacturers – formed a 50:50 joint venture company that is now the leading supplier of natural gas engines for buses around the world.

Cummins Westport Inc. (CWI[5]), co-located with Westport's corporate offices in Vancouver, Canada, undertook a very broad market and product development program that looked at Cummins markets worldwide, and concluded that the best opportunity for CWI was in focusing on the global market for mid-range engines (roughly 150hp to 300hp) used in urban transit buses and urban work trucks such as refuse collection vehicles.

Cummins manufactures and distributes the natural gas engines that are designed and marketed by CWI, keeping costs low and maximizing efficiency. CWI now has more than 12,000 natural gas engines in service in fleets in 28 countries, including China. Its most important global customer is Beijing Public Transit, which operates one of the largest advanced natural gas transit fleets in the world, with more than 2,500 buses powered by CWI engines. CWI has also sold engines to eight other Chinese cities, including Shanghai and Xian, and Chinese bus manufacturers have begun to export complete buses incorporating CWI engines. CWI engines are currently manufactured in North America and shipped to Chinese domestic bus manufacturers. In 2006, CWI plans to commence local China assembly of engines with Dong Feng Cummins as the production partner.

North American experience: Clean Energy Fuels Inc.

Around the same time as the formation of CWI, Westport partnered with Terasen (at that time named BC Gas, one of Canada's largest natural gas utilities) in a 60:40 venture to market natural gas as a vehicle fuel. Complete customer solutions require a fully-enabled supply chain for high-quality OEM vehicles, but also reliable and cost-effective fuel and fueling infrastructure.

Like the vehicle OEMs, many North American natural gas utilities entered the NGV industry in the 1990s. Typically the strategy was to build public-access fueling stations within the utility's geographic market area to serve the expected consumer passenger car markets. When this market did not materialize as hoped, most utilities abandoned their NGV business plan.

The strategy adopted by the Terasen/Westport partnership was different: market to large "anchor" fleets such as transit and refuse haulers in each city to ensure acceptable business economics, and enable consumer fuel stations only where it makes economic sense. In particular, Westport encouraged the team to pursue the newly-emerging market for LNG as a vehicle fuel. Joint marketing with CWI's team – to provide complete solutions to fleets, with the best vehicles and the best fueling support – has proved to make strategic sense for both companies.

This strategy has worked very successfully. Through a sequence of mergers and acquisitions, Clean Energy Fuels Inc. is now the largest natural gas fuel company in North America, with customers from Boston to San Diego.

Case study: benefits of large scale deployment in highway trucks

In order to illustrate the potential economic and environmental benefits of adopting LNG as a vehicle fuel in the heavy-duty sector, let us model the impact on just the heavy-duty highway truck market in China.

Assumptions:

- RMB 4 per liter of diesel fuel; RMB 2.15 per diesel liter equivalent LNG (delivered)
- 80,000 liters consumption per year delivers savings per truck of RMB 127,000 annually
- Heavy-duty trucks represent approximately 9% of the commercial vehicle population in China
- Approximately 5,000 LNG trucks sold in 2009 rising to 50% of new sales in 2020
- Average diesel population is operating at Euro II standards through the period.

With these assumptions, about 1.3 million LNG trucks would be on the road in 2020.

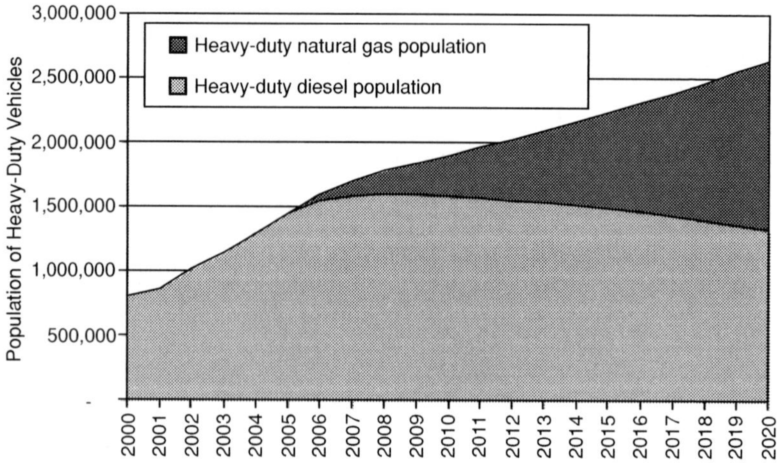

Source: Adapted from World Markets Research Centre and Westport Analysis

Figure 11.4 Growth of LNG Trucks to 2020

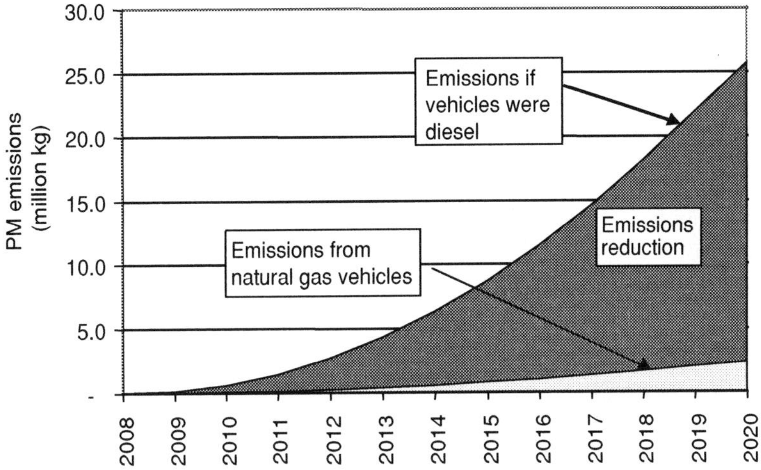

Source: Adapted from World Markets Research Centre and Westport Analysis

Figure 11.5 Modeled Particulate Matter (PM) Emissions Reduction

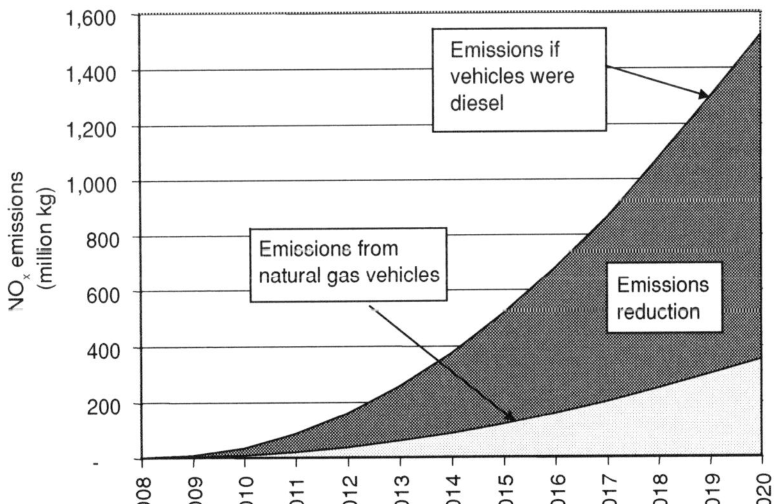

Source: Adapted from World Markets Research Centre and Westport Analysis

Figure 11.6 Modeled NO_x Emissions Reduction

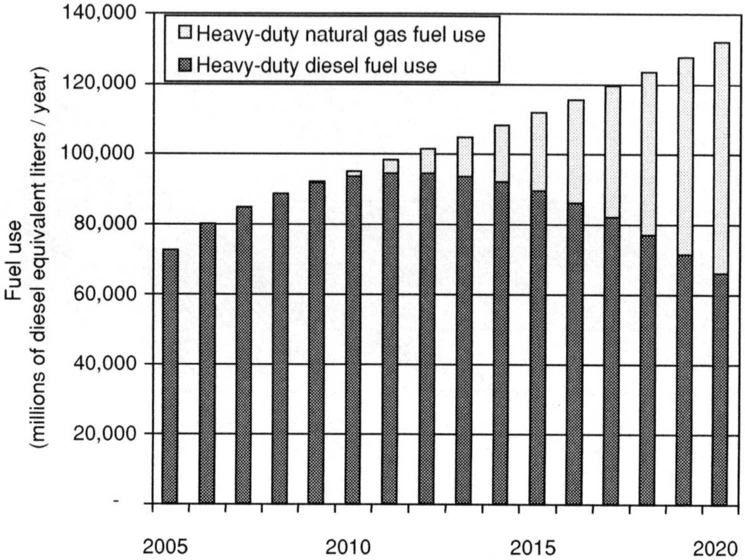

Source: Adapted from World Markets Research Centre and Westport Analysis

Figure 11.7 Modeled Oil Displacement

Emissions benefits are striking: more than 1.2 billion kg of NO_X and 23 billion kg of particulate matter would be avoided. Even more striking is the oil displacement achieved: over the period, more than 300 billion liters of diesel fuel, or more than 14% of China's projected oil imports during the period. The fuel cost savings at the modeled price would be RMB 550 billion – approximately US$100 billion.

Conclusion

Westport is confident that China will carry through on its stated goal to increase natural gas use from the present 2% of its total energy supply to 10% by 2020 in order to displace oil imports, to reduce air pollution, and to reduce dependency on insecure energy sources. This shift will include an acceleration of natural gas as a primary fuel for transportation. Westport also foresees that China will become the world leader in clean transportation within the next decade, and will use this to create leverage for its inevitable entrance into the global automotive market with differentiated technology and products.

Notes

1. IEA, *Energy Outlook*, Projected Energy Use in China to 2030, International Energy Agency, Paris, 2005.
2. IEA, *Energy Outlook*, China's Vehicle Population Growth, International Energy Agency, Paris, 2004.
3. See <www.acidrain.org/pages/publications/factsheet/factsheet17.pdf> for a good introduction to the regulations and potential emissions abatement technologies.
4. IEA, *Developing China's Natural Gas Market,* International Energy Agency, Paris, 2002.
5. See <www.cumminswestport.com>.

12
Challenges on the Pathway to a Hydrogen Economy

Karen Hall

Hydrogen energy technologies are commercially available today in niche applications where hydrogen's unique benefits bring special value. The National Hydrogen Association (NHA) believes that the path to commercialization of hydrogen energy systems will only be successful if industry and government work together in partnership. Neither group can succeed alone.

The chapter will detail the NHA's view on the following:

- Transitioning to fuel cells
- Systems analysis coordination
- Hydrogen use by the federal government
- Safety, handling and utilization
- Coordination of federal hydrogen efforts
- Important role of education and information dissemination
- Need for demonstration projects.

In the coming decade, the NHA will continue to share its vision of a world powered by clean, inexhaustible energy sources – through hydrogen. This vision, which represents a commitment by the NHA's member organizations can help shape national and international energy policy, thus paving the way to a global hydrogen economy.

Introduction

In 2003, the National Hydrogen Association (USA) developed and adopted *On the Road to Hydrogen: Policy Priorities*. The NHA is an industry trade association of more than 100 companies and organizations whose mission is to foster the development and use of hydrogen technologies and to promote the use of hydrogen as an energy carrier in the United States. Membership includes many international and multinational organizations.

This chapter presents the NHA's Policy Priorities document, which addresses needs, pathways and policies that the NHA feels are important to meeting its mission. *On the Road to Hydrogen: Policy Priorities* describes the consensus view of the context, issues and challenges facing the USA in the development of a hydrogen economy.

It is worth noting that while other countries and regions may have differing primary drivers for working toward a hydrogen energy economy, the issues and challenges remain the same. Many countries have created or are in the process of creating a national hydrogen association, and many are working internationally through broader organizations such as the European Hydrogen Association, the Partnership for Advancing the Transition to Hydrogen, the International Energy Agency, and the International Partnership for the Hydrogen Economy.

Drivers to hydrogen

The primary drivers in the USA today for moving toward a hydrogen energy future are:

- National energy security
- Environmental stewardship
- Economic prosperity.

The US oil gap continues to grow. The *Transportation Energy Data Book, Edition 21 (Sept. 2001)*, and the *EIA Annual Energy Outlook (2002)* project declining US oil production, and increasing demand. Projections based on current trends indicate that as much as 70% of the oil consumed by the USA will come from imports by 2025. Hydrogen made from a broad array of domestically available sources will help to diversify our primary energy sources. Distributed energy resources, such as turbines, advanced gensets and fuel cells, provide opportunities to meet increased power needs without increased transmission infrastructure.

These systems can operate independently of the grid and when they are operated on hydrogen or hydrogen-blended fuels or both, significant reductions in urban air pollution and carbon dioxide emissions will be realized. The hydrogen-powered car also offers the possibility for emissions-free and greenhouse gas-free transportation.

Hydrogen offers a strategic opportunity to meet increasing US energy needs in a highly efficient way without increasing greenhouse gas emissions. A transition to hydrogen energy systems will facilitate economic growth by reducing our dependence on oil from volatile regions, by cutting our foreign trade balance, by decreasing the magnitude of oil price swings while at the same time increasing domestic economic activity. The USA has a leadership position in hydrogen manufacturing capability and in advanced hydrogen technology development, which can be leveraged into a continuing worldwide leadership position.

The USA is well positioned for a transition from oil and gas to hydrogen as the primary energy carrier. Hydrogen, along with electricity, will then be the principal energy carrier, used in all sectors of society and generated from all available feed stocks. The members of the NHA are among the major industries dedicated to creating a hydrogen energy infrastructure, including several major multinational energy suppliers. The NHA, drawing on technical expertise from members including NASA and the industrial gas suppliers with decades of experience supplying and using hydrogen safely, is working hard on developing the safety codes and standards that will enable the hydrogen infrastructure.

The role of the NHA

Transition to a future hydrogen energy economy requires commitment from many stakeholders, including energy companies, industrial gas suppliers, automotive manufacturers, technology manufacturers, government organizations, universities and research organizations.

The membership of the NHA encompasses each of these stakeholders, which have made a commitment to hydrogen. The NHA's strength stems from the diversity of its membership and draws upon the expertise of these leading organizations in hydrogen safety, production, storage, distribution and utilization. This diverse membership enables the NHA to maintain objectivity in our role and commitment to a transition to a hydrogen economy.

The NHA is a natural forum for building consensus, developing viable pathways to a hydrogen economy, acting as a catalyst in the development of consensus codes and standards, advocating partnerships, advising government agencies on policies and demonstrations, offering guidance on research and development (R&D) programs to achieve certain technical objectives, and serving as a clearing house for educational information.

NHA commitment and proposition

The members of the NHA propose that Congress and the Administration implement a much broader and much more intense Hydrogen Program than exists today. The goal would be to support private sector development and demonstration of every aspect of hydrogen infrastructure and use. The NHA will work with the government and other groups to organize and coordinate the hydrogen program team. We can act as a point of contact and central facilitator to develop teams to carry out the activities of this program. We can host meetings and workshops to provide a comprehensive view of the technologies and challenges. We can also work on an educational level, to explain the value and importance of the activities in the program to policymakers, partnering agencies and American citizens.

In the coming decade, the NHA will continue to share its vision of a world powered by clean, inexhaustible energy sources – through hydrogen. This vision, which represents a commitment by the organizations represented by the NHA, can help shape national and international energy policy, thus paving the way to a global hydrogen economy.

Transition to a hydrogen future

The NHA recognizes the important role of domestic fossil fuel resources including coal, gasoline, natural gas and diesel in the transition to a hydrogen economy. In the near term, a significant portion of hydrogen will be produced from fossil fuel feedstocks. The NHA supports further technological development for these processes, aimed at reducing the cost of hydrogen and reducing or eliminating the environmental effects from these methods.

The NHA also supports technological and economic development of renewable energy technologies, and envisions a growing portion of future hydrogen production to come from renewables. The end point would be a diverse portfolio of hydrogen generation technologies and feedstocks, with as much hydrogen production from renewables as is practical. Longer-term continuation of fossil fuel feedstock and even nuclear production of hydrogen are possible if the environmental damage and security issues from these methods can be eliminated or sufficiently mitigated. Ultimately, society and the market will determine the energy mix consistent with resources, social structure, population, economy and environmental requirements.

Production and uses

Technologies are available today for the safe production, storage and utilization of hydrogen energy. There are, however, certain technical breakthroughs necessary if hydrogen is to live up to its full potential. These breakthroughs would enable hydrogen energy to compete economically with technologies operating on conventional fuels. Breakthrough areas include:

- Efficient, economical, renewable energy production
- High density, on-board storage technologies for transportation applications
- Higher efficiencies in electrolysis of water
- Technologies to reduce the environmental effects of other methods for hydrogen production, such as carbon sequestration from fossil fuels, and hazardous materials and security issues associated with nuclear energy.

Hydrogen energy technologies are not limited to fuel cells. They also include reciprocating internal combustion engines, turbines, electrolyzers, and other technologies that use or produce hydrogen. Hydrogen's unique properties allow for more efficient and environmentally benign application of conventional technologies as well as enabling advanced emerging fuel cell technologies.

Applications for hydrogen energy systems include:

Portable power

Portable power provides significant opportunities for early product introductions into smaller markets. Portable power applications generally tolerate higher energy costs for convenience. Portable markets will decrease the cost of hydrogen technologies for other applications.

Stationary/distributed power

Hydrogen can play a role in the generation of electricity on-site, at or near the point of use and at fixed power plants. Distributed power applications provide near-term market opportunities. Stationary power markets are mid-term opportunities for hydrogen and offer increased efficiencies to the grid. Stationary power provides the largest opportunity for CO_2 reductions, along with the reduction or elimination of other greenhouse gases. This is especially true if the stationary power generator is used for combined heat and power (CHP). CHP systems have the potential to substantially increase the efficiency of distributed power generators.

In remote locations without an electrical grid, electricity from solar, wind, tidal and other renewable sources can be used increasingly in a sustainable world. These sources of electricity are often intermittent and need an energy storage system to meet peak power requirements. Hydrogen produced from off-peak or surplus power can be used to store energy for delivery – as electricity – when needed. Hydrogen can be used to supply electricity to each home and business by locally sited power plants, such as hydrogen-fueled microturbines, reciprocating engines, or fuel cells, which provide the electricity as well as space heating, with no air or water pollution, in a quiet, unobtrusive manner.

Transportation fuel

There is strong and growing interest in using hydrogen as a transportation fuel. With the market price of transportation fuels being higher than the market price for other applications this offers a unique opportunity for hydrogen to become cost competitive with conventional fuels.

The NHA believes a thrust in the area of transportation will provide the largest, long-term opportunity for commercial application of hydrogen energy technologies and contribute to the creation of a hydrogen energy infrastructure. Automobiles provide the best opportunity to engage the public now in the benefits and reasons to move toward hydrogen energy. Buses and fleets, however, can provide an even earlier market, with fewer infrastructure considerations through use of centralized refueling and should be a central part of near-term programs.

Commercialization

Hydrogen energy technologies are commercially available today in niche applications, such as space propulsion, power systems and industrial applications where their unique benefits bring special value. However, using these technologies in early products employing hydrogen as an energy carrier requires economic support to make them commercially viable and affordable.

The NHA believes that the path to commercialization of hydrogen energy systems will be most successful if industry and government work together in partnership. Neither group can succeed alone. Both have unique strengths.

Industry has the expertise and financial resources to bring new products to the marketplace. But industry must respect the bottom line and the demands of its stockholders to provide short-term return on investment. It cannot finance long-term societal goals such as clean air and reduced dependence on foreign oil.

Conversely, governments do have the resources and responsibility to encourage new technology for the marketplace. Governments have the sole charter to "protect the commons" – in this case, energy security and the environment. Government alone has the long-term staying power and the mission, acting on behalf of all citizens collectively, to develop and promote new technology and new policies that will achieve societal objectives. Working together, government can provide the seed money and regulatory environment to start the process, and industry can provide market knowledge and the large-scale capital investment, once the hydrogen technology development comes within industrial planning horizon time scales.

Economically competitive applications are needed for fuel cells to meet cost and performance targets. When fuel cell technology is mature, it will be able to produce high value power, to increase reliability, power transportation vehicles, and improve communications with applications such as relay stations, personal computers and cellular telephones. Most importantly, fuel cells offer the possibility of higher energy efficiencies than conventional technologies.

A transition to the hydrogen energy future can begin prior to high volume commercial adoption of the fuel cell. Hydrogen can be used today in fuel cells for back-up power applications, reciprocating internal combustion engines for vehicles, as well as turbines and microturbines.

Exploiting the unique properties of hydrogen, conventional devices fueled on hydrogen can be designed to be zero or near-zero-emission technologies. The near-term commercialization of hydrogen internal combustion engine vehicles and hydrogen/natural gas or hydrogen/electric hybrids allows a platform for hydrogen energy technology commercialization. This can facilitate public acceptance of hydrogen and provide the beginning of a hydrogen infrastructure. As fuel cell technology development continues, it can find its place faster with a hydrogen infrastructure in place.

Critical pathways

In addition to the technological breakthroughs mentioned above the NHA believes that other critical pathways are necessary to create knowledge to meet the objectives of transitioning to a hydrogen energy future. These are:

- *Systems analysis coordination.* Both the automotive and fuel supply industries have to be concerned with their bottom lines as well as with the environment or oil imports, so the government needs to separately evaluate the societal impact of these hydrogen transportation market penetration scenarios. Hence the need for a coordinated systems analysis encompassing all three entities: vehicles, fuel supply, and society.

- *Hydrogen use by the federal government*. The federal government can and should procure hydrogen energy systems for its own needs. This includes fleet vehicles with hydrogen refueling as well as stationary and distributed power systems for buildings and mobile power applications. Federal use of hydrogen energy technologies would provide high visibility for these technologies and an opportunity to educate the public about the benefits of hydrogen.
- *Safety, handling and utilization*. Research, development and demonstration (RD&D) should be conducted for safety, handling and utilization of hydrogen fuel by the public and these lessons must be incorporated into the developing codes, standards and regulations. Safety must be a consideration across the board. The need for safety-related activities is urgent, which is why the NHA is leading the timely development of codes and standards. Participation by permitting agencies and the insurance industry is urged, along with equipment testing programs. The federal government must provide implementation training and barrier reduction downward to local and regional government. The NHA will work with government to develop programs for training and barrier reduction.
- *Coordination of federal hydrogen efforts*. To focus and coordinate federal hydrogen efforts, creation of an Interagency Hydrogen Task Force is urged, with participation by the US Departments of Energy, Transportation, Defense, and Commerce; the US Environmental Protection Agency; the Department of Homeland Security; the National Aeronautics and Space Administration; the National Science Foundation; and the National Park Service. The NHA encourages an Administration mandate to promote the use of hydrogen within agencies in order to assist in the development of hydrogen-related business.
- *Important role of education and information dissemination*. At the international, national, state and local levels, efforts should be initiated to encourage strong community programs to develop, to serve market development needs, and to increase the understanding of all stakeholders. Databases and previous research results should be made available as broadly as possible to promote the transfer

of knowledge, broaden research experience, and introduce new ways of thinking to established research organizations. It should be recognized that education is a continuum that stretches from creating knowledge to educating in schools, training workers, educating policymakers and informing the public.

- *Need for demonstration projects.* The NHA supports demonstration activities to achieve the goals of transitioning towards a hydrogen economy in stationary and transportation markets and to help understand and develop the hydrogen infrastructure. Demonstration projects, jointly funded by government and industry partnerships, should begin as soon as is possible and practical and should be operational in a variety of applications and environments across the United States. The NHA supports a full and open competition that encourages partnerships, results in a wide range of demonstration parameters, requires safety analysis, and provides a broad array of high-profile educational and information dissemination opportunities.

Hydrogen policy

The NHA applauds the Department of Energy's National Hydrogen Energy Roadmap. The Roadmap, developed in 2002, is a well-balanced plan with an intelligent transition strategy, first relying on conventional feedstocks and optimized hydrogen-fueled conventional conversion devices to pave the way for the introduction of the fuel cell.

The NHA supports the President's Hydrogen Fuel Initiative, announced in his State of the Union address on January 28, 2005. This initiative will go a long way toward creating the infrastructure necessary for clean transportation using domestically-produced hydrogen.

The NHA recognizes the need for economic incentives, including tax policies, at the appropriate point in the technology development and early commercialization. We advocate increasing incentives as technologies – such as hydrogen-powered near-zero emission vehicles and zero-emission vehicles (ZEVs) – become available for buyers. We also advocate incentives (rather than mandates, requirements or regulations) to accelerate market penetration. This includes voluntary emission credit trading schemes to begin to manage greenhouse gas emissions.

The NHA also supports economic incentives and initiatives involving the use of hydrogen for stationary power, portable power, and transportation. The organization pledges to work to make all these visions a reality.

13
How the Clean Air Interstate Rule Will Affect Investment and Management Decisions in the US Electricity Sector

Caryl Pfeiffer

This chapter explores the impact of the Clean Air Interstate Rule (CAIR) on the regulated utilities of LG&E Energy and its regional competitors. The way in which these tightened air emission standards impact any individual utility in the USA will depend on each company's generation mix and regulatory status.

For LG&E Energy, its resultant investment will mean a 13% rise in revenue requirements for its retail customers or an increase of approximately $7 per month for a typical residential customer using 1,000 kilowatt-hours per month. For its competitors, such as the Tennessee Valley Authority (TVA) and American Electric Power (AEP), the effect will be to increase their rates; for both TVA and AEP by an additional 16%. Since LG&E Energy's rates will continue to be lower than both TVA and AEP, even after compliance with CAIR, LG&E Energy will maintain its position as a cost leader. This is because it has not only a lower installed cost for its flue gas desulfurization (FGD) systems, but also fewer coal-fired megawatts that needed to be retrofitted with SO_2 control equipment.

Introduction: Overview of the Clean Air Act Implementation Rule

The Clean Air Act (CAA) of 1970 (amended in 1977 and in 1990) is the US framework for establishing National Ambient Air Quality Standards (NAAQS) to protect public health and welfare and to control harmful emissions from sources of air pollution in order to achieve those standards. The Clean Air Act is administered by a federal agency, the US Environmental Protection Agency (EPA).

In July 1997, the EPA revised the National Ambient Air Quality Standards for ozone and particulate matter. The new NAAQS for ozone changed the existing standard from one that measures ozone concentrations over a one-hour period to one that measures over an eight-hour period, and tightened the concentration of the pollutant from 120 parts per billion (ppb) to 80 ppb. For particulate matter less than 2.5 microns in diameter ($PM_{2.5}$), the EPA identified NO_x emissions from fossil-fired power plants as a major contributor to ozone pollution, and acid aerosols (NO_x and SO_2) from fossil-fired power plants as fine particulates likely to be subject to additional controls.

The EPA designated eight-hour ozone and $PM_{2.5}$ nonattainment areas (counties) based on ground level monitoring of pollutant levels. If an area's monitored concentration of a pollutant exceeds the standard, then EPA categorizes that as nonattainment. Each state with a nonattainment area, or areas, then submits a plan for attainment (a State Implementation Plan or SIP).

The EPA determined that attainment of these new NAAQS in the Eastern United States could only be achieved through a combination of emission reductions from:

1. sources located in or near nonattainment areas (such as mobile sources); and,
2. sources (such as fossil-fired power plants) located further from the nonattainment areas to deal with interstate transport of air pollution.

To address the regional transport of air pollution and its "significant" contribution to ozone and fine particulate matter nonattainment in downwind areas, the EPA promulgated the Clean Air Interstate Rule (CAIR), finalized on March 10, 2005. This requires significant SO_2 and/or NO_x emissions reductions from fossil-fuel fired power plants in 28 Eastern states and the District of Columbia.

The rules are intended to reduce further ground level ozone (by requiring further reductions in NO_x during the summer) and to reduce fine particulates or $PM_{2.5}$ (by reducing SO_2 and NO_x on an annual basis).

It requires the following reductions from fossil-fired electric generating units (EGUs) compared to their 2003 levels – see Table 13.1 – and imposes the following caps of emissions – see Table 13.2.

Table 13.1 Reductions of Fossil-Fired EGUs

Applies during:	28 States		Kentucky
Annual NO_x	2009–14	53%	42%
Annual NO_x	2015 and beyond	61%	58%
Annual SO_2	2010–14	45%	36%
Annual SO_2	2015 and beyond	57%	49%

Table 13.2 SO_2 and NO_x Emissions Caps

	SO_2	NO_x
2004	7.9 million tons	2.7 million tons
2011	3.6 million tons	1.5 million tons
2015	2.5 million tons	1.3 million tons

Ozone Season NO_x: replaces the current NO_x SIP Call (which caps NO_x emissions during the May–September ozone season) with CAIR NO_x caps, also during May–September. For Kentucky, the new cap is identical to the NO_x SIP Call for 2009–14 and is reduced by about 15% for 2015 and beyond. Therefore, beginning in 2009, there will be two overlapping restrictions on NO_x emissions, and EGUs must comply with both. They must emit less than a given amount during the ozone season and emit less than another given amount on an annual basis.

The rule requires states to submit a plan to EPA on how they will achieve the reductions, either by participating in a regional "cap-and-trade program" (similar to the Acid Rain and NO_x SIP Call programs) or by an alternative of the states' choosing. The plans were to be submitted by September 2006.

Note on acid rain cap and trade program

The Acid Deposition Control Program was established under Title IV of the Clean Air Act Amendments of 1990. Title IV's purpose was to reduce the adverse effects of acid deposition through a permanent 10 million ton reduction in SO_2 emissions and a 2 million ton reduction in NO_x emissions from 1980 levels in the 48 contiguous states. The Acid Deposition Control Program established a cap on annual SO_2 emissions at approximately 8.9 million tons by the year 2000. The legislation obtains these SO_2 emission reductions from electric utility plants through the use of a market-based system of emission allowances. Under the system, "affected units" are allocated allowances in an amount based on their historic fossil fuel consumption and a specified emissions rate. Affected units are required to hold sufficient allowances to cover their annual level of SO_2 emissions, but are given the flexibility to reduce their emissions in the most cost-effective manner. Once allocated, allowances may be used by affected units to cover SO_2 emissions, banked for future use, or sold to others. SO_2 allowances are not unit-specific and can be sold or traded. Utilities have incentives to reduce SO_2 emissions so that they can use allowances to cover growth, to make additional off-system sales, or to sell or trade allowances on the open market, offsetting the cost of compliance. Utilities that reduce the amount of SO_2 emitted below their authorized level may bank their allowances for future use or sell or trade them to other utilities that expect to exceed their authorized emission level or to other allowance market participants such as emission brokers.

Note on NO_x SIP Call

In September 1998, EPA finalized the NO_x SIP Call rule, requiring 19 Eastern states, including Kentucky, and the District of Columbia to revise their State Implementation Plans. The new SIPs must achieve additional NO_x emissions reductions that the EPA believes are necessary to mitigate the transport of ozone across the Eastern half of the United States. The final rule required fossil-fired power plants in the 19-state area to retrofit their electric generating units with NO_x control devices by May 2004. It is likely that Kentucky and most of the other affected states will choose the cap-and-trade approach under CAIR. Under a cap-and-trade program, each combustion unit is awarded a set number of "allowances." Historically, the unit would surrender allowances in an amount equal to its emissions to be in compliance; the new rule modifies the surrender ratio for SO_2, explained later. Each allowance has an associated vintage year and cannot be used for compliance before its vintage.[1] Allowances can be traded between units, plants, companies, and so on.

The allowance programs will be complicated since different states are subject to different combinations of requirements under the Acid Rain Program, the NO_x SIP Call, CAIR ozone-season reductions, and CAIR SO_2/NO_x annual reductions.

The following is a simplified summary based on the EPA's model emissions trading rule:

SO_2

Existing Acid Rain allowances will be used by LG&E Energy. Allowances with vintage 2009 and earlier would be surrendered on a "one-for-one" basis throughout the CAIR program. Vintages 2010 through 2014 would be surrendered on a "two-for-one" basis (surrender two allowances for each ton of emissions) and vintages 2015 and beyond would be surrendered on a "2.86-for-one" basis. This acts as an incentive to reduce emissions and bank SO_2 allowances before 2010.[2]

NO_x

LG&E Energy's allocation will not be known until the state of Kentucky develops an in-state allocation process. A range of estimates has been provided below:

The EPA will allocate predetermined numbers of NO_x allowances to each state and the individual states will determine how to allocate these to individual units, similar to the current process under the NO_x SIP Call. Because Kentucky is required to reduce NO_x for both ozone and $PM_{2.5}$ there will be ozone season allowances and annual allowances.

For ozone-season control, Kentucky's allocation is the same as under the NO_x SIP Call for 2009–14. For 2015 and beyond, its ozone season cap is about 15% lower (nominally based on 0.125 lb/mmBtu vs 0.15 lb/mmBtu).

For $PM_{2.5}$ control, Kentucky's annual NO_x allocation is 83,205 tons during 2009–14 and 69,337 tons for 2015 and beyond. These figures are about 7% higher than in the proposed regulations, largely because the EPA applied a weighting factor that allocates more allowances to coal-fired generation than to oil and gas; thus, Kentucky's allocation increased because of its high percentage of coal-fired generation. A ballpark estimate of LG&E Energy's possible allocation is provided in Table 13.3. However, Kentucky may choose to set some allowances aside for new sources or to withhold some and auction them. For example, under the NO_x SIP Call, Kentucky withheld 5% of the 2004–06 allocation and auctioned them, with proceeds going to the Kentucky General Fund.

Table 13.3 Estimate of Annual NO_x Allowances (assuming no withholding)

	2009–14	2015 and beyond
KU	16,300	13,400
LG&E	12,600	10,500

The EPA determined that the most "cost effective" way to attain the new ground level NAAQS for ozone and fine particulate matter was with emission reductions from fossil fuel-fired boilers. The EPA also calculated that CAIR costs are "modest in light of benefits expected."

The EPA estimates that CAIR will result in substantial human health and environmental net benefits (up to $100 billion per year) while helping urban areas in the East to achieve the new, more stringent NAAQS for ozone and fine particles. The cost of the program is estimated at $3.6 billion per year (the investment required by the electric utility industry in the East). Since these costs will fall upon the electricity consumers, the EPA estimates that electricity prices will increase 2.0–2.7% within the CAIR region in the 2010–15 timeframe.

Regulatory drivers leading to CAIR

The principal drivers leading to the introduction of CAIR are:

1) *Emission reductions*. In the US, there is an ongoing push by public health and environmental advocates, as well as by regulatory agencies, for emission reductions from coal-fired generation. There is 244 GW of installed coal-fired capacity in the CAIR footprint, which represents 80% of the coal-fired capacity in the United States. The electric generation power sector represents 69% of the total SO_2 emissions and 22% of the total NO_x emissions in the United States.
2) *Local governments' inability to regulate mobile sources*. Even though mobile sources account for the overwhelming majority of emissions resulting in ground level ozone in urban areas, local elected officials have been unwilling to impose additional restrictions on passenger vehicles because the passengers in those vehicles vote.
3) *Need for local governments to get reclassification of their counties to attainment status*. If a nonattainment area is not on an EPA-approved State Implementation Plan to attain the NAAQS, federal government funding for transportation projects is halted. If local governments adopt (through their SIPs) the EPA regional solution that only affects coal-fired power plants (not mobile sources), then they have an EPA-approved plan without making hard choices about local emission reductions.
4) *Disparity of electric prices in the Eastern United States*. East Coast electricity producers have not achieved the lower rates promised through deregulation and are looking for ways to increase their competitive edge. Some East Coast utilities are hoping that rate equalization will occur as environmental compliance costs are pushed onto the heavily coal-dependent (and lower cost) electric

generators in the Midwest and South. Currently, East Coast electric providers charge an average retail rate of 12¢ per kilowatt-hour (kWh), as compared to providers in the Midwest and South who come in, on average, at 7.7¢ per kWh.

LG&E Energy's current situation

LG&E Energy is a coal-fired electric utility located in Kentucky. It has 9,005 megawatts (MW) of installed capacity of which 8,190 MW is coal-fired. Its 27 coal-fired units, at ten sites throughout Kentucky, burn 21 million tons of coal per year. As shown in Figure 13.1, 98% of the kWh generated and sold comes from coal-fired units.

LG&E Energy's position in the marketplace is based on three main factors:

1. Cost – LG&E Energy has the lowest electric rates in the United States, due to:

   - a low embedded cost of generation
   - the location of power plants adjacent to low-cost coals (the Illinois Basin, the Appalachian Basin and Pittsburgh 8)
   - its historic position as leader in air pollution control – LG&E Energy installed the first flue gas desulfurization system (FGD) in the US (1974), enabling the company to gain experience with this control equipment while burning higher sulfur, lower cost coals. The company also invested early in selective catalytic reduction systems and has demonstrated among the highest NO_x removal rates for this technology in the US
   - 60% of the company's installed coal-fired capacity is being equipped with FGD for SO_2 control
   - 84% of its installed coal-fired capacity is being equipped with selective catalytic reduction (SCR) for NO_x control
   - good utility management and fair regulatory agency oversight.

2. Reliability – LG&E Energy has:

   - adequate reserve margin in generating capacity
   - a robust transmission system across the service territory, although there are certain constraints in extreme Western Kentucky under peak flow conditions (not caused by LG&E Energy's system, but by those we are interconnected to – TVA and BREC).

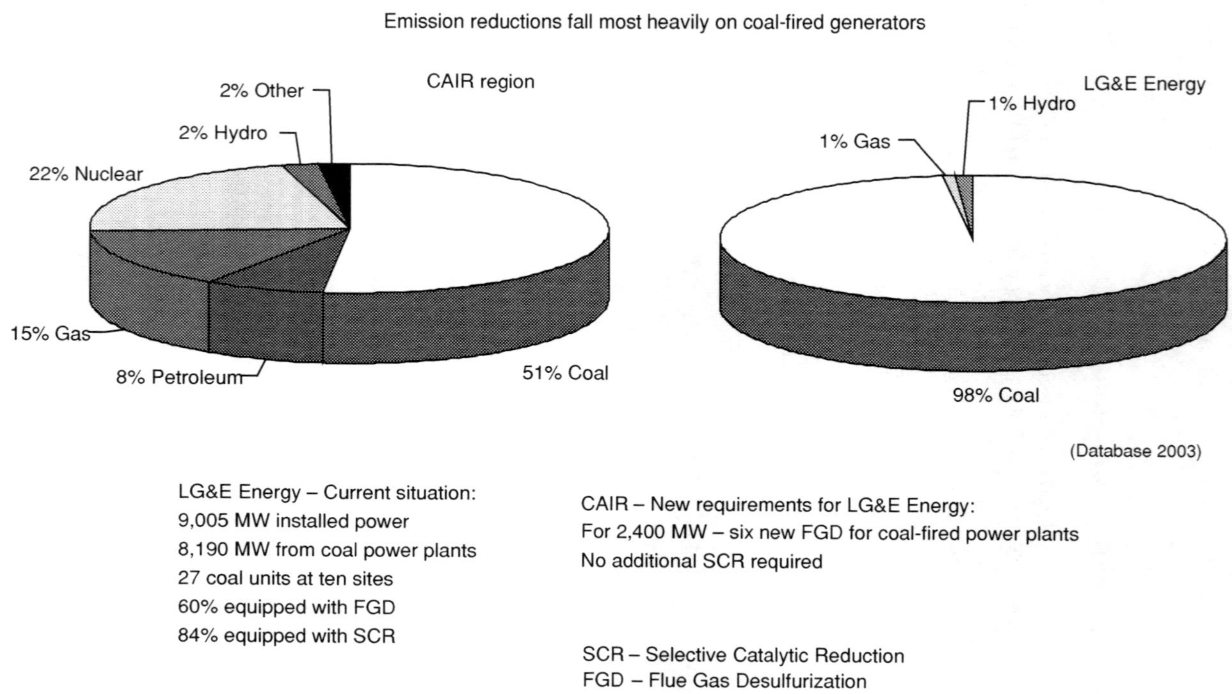

Source: LG&E Energy

Figure 13.1 Electricity Generated by Energy Sources

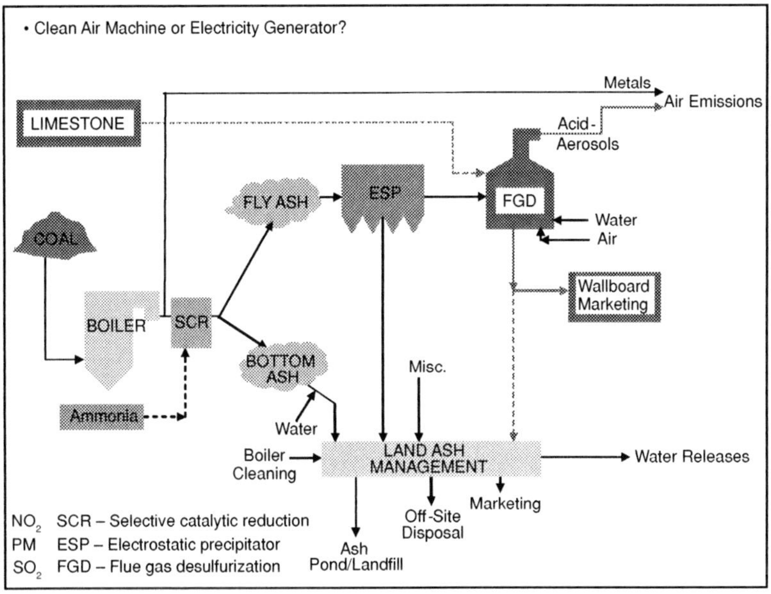

Source: LG&E Energy

Figure 13.2 Air Pollution Control Technologies for Coal-Fired Units

3. Customer service:

  - LG&E Energy has been awarded the JD Powers customer service award (for both residential and commercial customers) for the past six years running
  - the company has highly qualified and dedicated customer service representatives for each of the customer classes – wholesale, industrial, commercial and residential.

CAIR compliance options for coal-fired electric generating units

The principal choices facing LG&E Energy are:

1. Invest in air pollution control technologies or lower emitting generation sources
2. Fuel switch, and/or
3. Rely on the emissions trading market.

Note that non-compliance is not an option. There are civil penalties of $27,500 per violation, and criminal penalties, where a utility's designated representative under the Clean Air Act is held personally liable, including the potential for jail time.

LG&E Energy CAIR compliance strategy

In response to the proposed CAIR rules, LG&E Energy undertook an assessment to determine the most cost-effective means of meeting these environmental requirements (within the emission limitations and deadlines imposed by the rule) and to provide the time frame for construction and the estimated cost of the projects. This is contained in the 2004 SO_2 Compliance Strategy study, which is a result of the company's continual environmental compliance analysis review process. As identified in our 2002 SO_2 Environmental Compliance Analysis (contained in the company's 2002 Integrated Resource Plan), a rapidly depleting bank of SO_2 allowances will soon require the addition of new control technologies or the purchase of SO_2 allowances. The company's SO_2 allowance bank reached a peak of 297,000 in 1999 and has diminished substantially since then due to load growth and the resulting increased utilization of the coal-fired generating units. By year-end 2004, the bank is projected to be around 160,000 allowances and it is likely to be completely depleted before the end of 2007. The shortfall that begins in 2007 will be further exacerbated by the implementation of the Clean Air Interstate Rule in 2010. The study identified that unless additional SO_2 reductions are achieved, LG&E Energy will be forced to acquire SO_2 allowances from the allowance market, where prices have rapidly escalated from approximately $175 a ton in July 2003 to nearly $700 a ton in November 2004 (and to over $900 in October 2005).

The 2004 SO_2 Compliance Strategy study identified that Ghent Units 2–4 and Brown Units 1–3 account for over 55% of the company's future SO_2 emissions and were the most logical and economical choice for installing control technologies to reduce overall SO_2 emissions. The study utilized the PROSYM™ detailed hourly production costing computer model from Henwood and the Strategist® Capital Expenditure and Recovery (CER) module from New Energy Associates to analyze 22 cases. The analysis included the evaluation of numerous strategies utilizing wet FGD processes, dry FGD processes, and fuel switching for these units, and the purchasing of allowances to determine the most cost-effective means of mitigating the expected SO_2 allowance shortfall. It explored the use of low-sulfur coals such as Powder River Basin and Eastern Compliance, but determined that fuel switching at both Ghent and Brown failed to significantly improve the overall SO_2 position. Sensitivities to capital cost and the market price of SO_2 allowances were also considered in the analysis.

The company's least cost plan results in the construction of wet-limestone, forced oxidation FGD systems on Brown Units 1, 2 and 3 and Ghent Units 2, 3 and 4. The anticipated costs and completion schedule is identified in Table 13.4. It is expected to take 18–24 months to complete each FGD, with some portions of the engineering and design being undertaken simultaneously.

Table 13.4 Flue Gas Desulfurization – Anticipated Costs and Completion Schedule

Location	In-service date	Total cost (million $)
Brown 1–3 FGD	May '09	234.19
Ghent 2 FGD	May '08	149.57
Ghent 3 FGD	May '07	129.02
Ghent 4 FGD	May '09	146.15
		658.93

NOTE: The SO_2 price forecast used in this analysis was a combination of actual market prices and forecast prices. The prices from 2005 through 2009 reflected the market prices at the end of June 2004 adjusted for inflation. The market prices at the time the forecast was established were the highest in the history of actively traded SO_2 allowances. The anticipated impact of CAIR on the allowance markets was reflected in the 2010 forecast allowance price. In the remaining years the prices were escalated based on the company's assumptions for inflation.

The supply side of the SO_2 allowance market has remained very conservative with the credits available, due to the projected impact of CAIR. These factors have contributed to the current high market prices for SO_2 allowances.

The 2004 SO_2 Compliance Strategy study recommends the construction of wet FGD systems on Ghent Units 2, 3 and 4 and Brown Units 1, 2 and 3, with the simultaneous switching of the units to high-sulfur coal and the purchasing of allowances on an as-needed basis as the most cost-effective plan for continued environmental compliance. While the addition of the FGD systems does not eliminate the need to purchase SO_2 allowances, the installation of controls significantly reduces the company's exposure to the SO_2 allowance market and further assures continued economical compliance with the environmental requirements of the Clean Air Act. The study also showed that, compared to purchasing SO_2 allowances, the construction of the wet FGD systems and the simultaneous conversion of the units to high-sulfur coal provided the following ratepayer benefits over the 20-year analysis period:

- decreases the cost of SO_2 compliance, with cost savings projected to be more than $110 million (PVRR[3]), and provides the most reasonable least-cost approach for continued environmental compliance;
- limits significant exposure to the SO_2 allowance market by reducing the company's anticipated shortfall;
- increases fuel procurement flexibility (lowers fuel costs); and
- improves position for meeting future regulations regarding mercury.

Note that under Kentucky state law, regulated utilities that retrofit air pollution control equipment in compliance with emission reduction requirements under the Clean Air Act, as amended, can apply for Environmental Cost Recovery (ECR) from the Kentucky Public Service Commission. This mechanism allows for the dynamic recovery of environmental control costs without the need to file a full rate case. Upon a demonstration that the utility's proposed Environmental Compliance Plan represents the least cost option for meeting the new requirements, the utility can recover the actual costs (plus a rate of return) associated with the capital investment in the air pollution control equipment. These incremental environmental compliance costs are recovered through an ECR bill surcharge.

Since the company's evaluation, there have been significant upward movements in SO_2 allowance market prices. This trend is positively correlated to the value that the company's compliance plan provides its ratepayers. As discussed in the 2004 SO_2 Compliance Strategy, a 10% increase in the SO_2 price forecast increases the value of the project by more than $54 million (from $110 million savings for the company's ratepayers to $164 million). As recently as October 15, 2005, SO_2 allowances were priced at $950 a ton, more than double the forecast 2005 price of $392 a ton used in this analysis. Therefore, the economics supporting the decision to construct wet scrubbers at Ghent and Brown would be enhanced by the incorporation of today's SO_2 market prices.

Impact on LG&E Energy and its competitors

The intention of CAIR is to force SO_2 and NO_x emission reductions from coal-fired generation in the Eastern United States. In order to achieve the SO_2 emission reductions imposed by the rule, LG&E Energy will install flue gas desulfurization on 2,400 MW of coal-fired generation (six electric generating units) at a capital cost of \$678 million. The project will decrease SO_2 compliance costs by \$110 million over the 20-year life of the investment and will result in a net present value of \$14.8 billion. Furthermore, LG&E Energy FGD retrofits will be installed at an average cost of \$275 per kilowatt. LG&E Energy is already in compliance with the NO_x emission reductions imposed by the rule, so no additional investment in SCRs will have to be made.

LG&E Energy's investment will result in a 13% rise in revenue requirements for its retail customers or an increase of approximately \$7 per month for a typical residential customer using 1,000 kWh per month.

LG&E Energy's major competitors (based on MWh sales and associated revenues) in the states that surround Kentucky are:

- Tennessee – Tennessee Valley Authority
- Missouri – Ameren
- Illinois – Excelon
- Indiana – Cinergy
- Ohio – American Electric Power
- West Virginia – American Electric Power
- Virginia – Dominion

Only TVA in Tennessee and AEP in West Virginia command rates today that could compete with LG&E Energy on "price" after LG&E realizes its rate increase due to compliance with the Clean Air Interstate Rule.

TVA has announced a \$1.5 billion investment in FGD on 4,350 MW to reduce SO_2 emissions (in addition to a \$1.3 billion investment in SCR on 9,875 MW to reduce NO_x emissions that it has recently completed) in order to comply with CAIR. The effect of this investment will be to raise their rates by an additional 16%. FGD retrofits will be installed at an average cost of \$340 per kilowatt.

AEP has announced a \$1.2 billion investment in FGD on 3,500 MW to reduce SO_2 emissions (in addition to an investment of more than \$1 billion in SCR to reduce NO_x emissions that it has recently completed). This, too, will raise their rates by an additional 16%. FGD retrofits will be installed at an average cost of \$343 per kilowatt.

Since LG&E Energy's rates will continue to be lower than both TVA's and AEP's, after compliance with CAIR, the company will maintain its position as a cost leader. This is due to LG&E Energy's having not only a lower installed cost for its FGD systems but also fewer coal-fired megawatts that needed to be retrofitted with SO_2 control equipment. LG&E Energy, as a regulated coal-fired generator, benefits from CAIR and will be able to not only recover the costs of compliance through the addition of air pollution control capital expenditures to its regulated rate base, but also have a resultant increase in regulated earnings.

Sources

CAIR information

Clean Air Act of 1970, as amended, 42 USC 7401 et seq., PL 91–604

Revisions to the National Ambient Air Quality Standards for Fine Particulate Matter and Ozone, 40 CFR Parts 50, 51 and 52, July 1997

Rule to Reduce Interstate Transport of Fine Particulate Matter and Ozone (Clean Air Interstate Rule); Revisions to Acid Rain Program; Revisions to NO_x SIP Call, 40 CFR Parts, 51, 72, 73, 74, 77, 78 and 96, March 2005

"A New Rule for Clean Air," *Electric Perspectives*, July/August 2004

Clean Air Interstate Rule, <www.epa.gov/cleanairinterstaterule>

Clean Air Interstate Rule, <www.epa.gov/cair>

Electric Utility Trade Association information, <www.eei.org/industry_issues/environment/air>

Emissions information

State-by-State Utility Emissions & Heat Input Data, Nationwide Utility Emissions & Heat Input Data, Emissions Scoreboard 2003, <www.epa.gov/airmarkets/emissions>

Energy supply, consumption and revenue information

Annual Energy Outlook, 100 Largest Electric Company Net Generation 2003, US Average Monthly Bill by Sector and State 2003, Net Generation by Fuel by State 2003, <www.eia.doe/cneaf/electricity>

Utilities: electricity, <www.worldmarketsanalysis.com/servlet>

LG&E Energy information

Assessment of Clean Air Interstate Rule Final Rule by Glenn Gibian, Senior Environmental Engineer, LG&E Energy Services, March 18, 2005

Control Technology Cost and Process Description information from Roger Medina, Senior Environmental Engineer, LG&E Energy Services

Testimonies of Robert M. Conroy and John P. Malloy in the Matter of: THE APPLICATION OF KENTUCKY UTILITIES COMPANY FOR A CERTIFICATE OF PUBLIC CONVENIENCE AND NECESSITY TO CONSTRUCT FLUE GAS DESULFURIZATION SYSTEMS AND APPROVAL OF ITS 2004 COMPLIANCE PLAN FOR RECOVERY BY ENVIRONMENTAL SURCHARGE, CASE NO. 2004–00426, <www.psc.state.ky.us>

SO_2 Compliance Strategy

Integrated Resource Plan

Rate Impacts of Least Cost Compliance Plan

Internal Investment Proposal to Construct Wet Limestone FGD Systems at Brown Units 1–3 and Ghent Units 2–4 and Fuel Switch to High Sulfur Coal, October 2004

Competitor information

Ameren: <www.ameren.com/corporateFacts> and <www.dynegy.com/Generation/Facilities/MAIN>

Excelon: <www.exceloncorp.com/generation/power>

Cinergy: <www.cinergy.com/sustain/environment-improvement.asp>

AEP: <www.aep.com/environmental/performance/emissionsassessment>

<www.aep.com/environmental/emissionscontrol/>

<www.aepchio.com/about/serviceTerritory>

<www.indianapower.com/about/serviceTerritory>

<www.kenntuckypower.com/about/serviceTerritory>

<www.apcocustomer.com/about/serviceTerritory>

Dominion: <www.dom.com/about/overview>

<www.dom.com/about/stations/fossil>

<www.dom.com/about/stations/hydro>

<www.dom.com/about/stations/nuclear>

TVA: <www.tva.gov/power/powerfacts>

<www.tva.com/environment/reports/envreports/aer2003/env_compliance>

Notes

1. Allowances cannot be surrendered to cover emissions earlier than the vintage year. For example, SO_2 allowances have already been awarded for 30 years into the future and have been placed by the federal government in individual accounts for each generating unit/company. When allowances are surrendered for operations in 2005, the company can use allowances dated with a vintage year of 2005 or earlier, but cannot pull allowances forward from a year in the

future to cover emissions at the present time. Hence, "No allowance can be used before its time – before its vintage year."

2. Increases in the exchange rates of allowances over time speed up the reduction of SO_2 emissions. In 2010 the exchange rate will be increased to 2 for 1, so that 2 allowances must be surrendered for every 1 ton of emissions into the air and in 2015 that goes to 3 for 1. So utilities are encouraged to reduce emissions because their allowances are worth less and less going into the future.
3. Present Value of Annual Revenue Requirements.

14
Future Investments in Power Generation: A Utility's Perspective

Sergio Castedo, Geir Holler and Bjørn N. Holsen

One of the first countries in the world to liberalize its electricity market, Norway is facing huge investments requirements in power generation to meet increasing demand and ensure security of supply. Evolving regulatory and technological environments offer a portfolio of options that utilities need to consider. This chapter looks at how the Nordic liberalization process shaped utilities' investment decisions and discusses Statkraft's investments in power generation in the short, medium, and long term.

Introduction

At the beginning of the 1990s, Norway was one of the first countries in the world to liberalize its electricity market in order to introduce competition. Generation, transmission and distribution were separated to improve the overall economic efficiency of the sector. Similar liberalization processes have followed in countries around the world with the aim of balancing supply and demand in an effective marketplace.

In the years to come, huge investments in new power generation will be required in order to meet increasing demand, replace the lost capacity of old generating stations and ensure security of supply. However, the ratification of the Kyoto Protocol and the continuous development of the renewable energy industry and policies have changed the portfolio of options that utilities consider for future investments in generation. The objective of this chapter is to present the perspectives of a major electric utility, Statkraft, with respect to past, current and future investments in power generation capacity in the Nordic power market. The chapter is divided into four sections. The first presents an introduction to Statkraft and the Nordic power market. The second reviews Nordic power generation investments since the liberalization of the electricity markets. The third considers today's investment climate in the power sector and addresses the principal trends and challenges. The fourth section presents Statkraft's view on future investments in power generation, and the chapter closes with some concluding comments.

Statkraft and the Nordic power market

The Norwegian power market was the first in the world to be opened up to general competition. The Energy Act of 1990, which divided the market into a competitive and a monopolistic sector, laid the foundation stone for deregulation in the other Nordic countries. In 1992, the major public company Statkraft was reorganized into the production and sales enterprise Statkraft AS, as it is known today, and the grid enterprise Statnett SF. With this division, the Norwegian state took the lead and pointed the way to more competitive and consumer-orientated power sales.

This section presents an introduction to the Statkraft Group, the Nordic power exchange (Nord Pool) and the main characteristics of electricity price formation in the Nordic area.

Statkraft

Statkraft is Norway's largest producer of electric power. It generates a total of 41.5 terrawatt-hours (TWh) per year, making it the third-largest producer of electricity in the Nordic region and the second-largest producer of renewable energy in Europe. It has an installed capacity of approximately 9 gigawatts (GW), comprising 91 hydropower plants, 113 water reservoirs, and two wind projects with a total installed capacity of 205 megawatts (MW).

The company's vision is to be a European leader in environment-friendly energy. Based on more than a century of knowledge and investment in the hydropower sector, the Statkraft Group is well equipped for further growth and development – and has a clearly stated desire to create lasting value. Statkraft stands for sustainable development in both an environmental and a financial sense. With approximately 2,000 employees including the Norwegian subsidiaries Skagerak Energi and Trondheim Energiverk, the group also has shareholdings in the Norwegian power companies BKK, Agder Energi and Fjordkraft, as well as in E.ON Sweden.

Statkraft is organized into three main business areas: (1) *Generation and markets*: responsible for the operation and maintenance of power plants, energy management, and trading and origination; (2) *New energy*: responsible for the development of new generation capacity, the generation of non-flexible assets, and innovation; and (3) *Regional:* responsible for developing Statkraft's domestic shareholdings and realizing synergies to increase profitability.

Its business concept is to become a leader in the construction and operation of environment-friendly power generation facilities, as well as in physical and financial power trading. The company seeks to maximize revenues from its production assets over time and uses dynamic hedging based on its own forecasts. Statkraft's definition of environment-friendly includes renewables and low-CO_2 emitting generation using state-of-the-art combined-cycle gas turbine (CCGT) technology.

Statkraft is working towards developing and participating in the following activities: sustainable energy production; the liberalization of energy markets; a secure power supply in Norway and the Nordic region; new capacity; NorNed cable; market-based solutions to dry-year problems; and a common central distribution grid operator for the Nordic region.

The company has the most flexible generation capacity and the largest reservoir capacity in Europe. In addition, it is active in financial and physical trading in the electricity and gas markets of continental Europe; an active participant on key European exchanges; responsible for operating the Baltic Cable; and active in cross-border trading across the continent. Finally, it is active in origination, structured products and asset optimization, renewable energy and green certificates, and CO_2.

Nord Pool

Nord Pool was the first power exchange in the world, and so far the only power exchange for trade across borders. It was established in 1993 as a wholly-owned subsidiary of the Norwegian national grid company Statnett. Prior to the creation of Nord Pool, Statkraft and Vattenfall controlled foreign trade. By January 1996, the Norwegian and Swedish markets were integrated. The Swedish grid company acquired 50% of the Statnett power market, which was renamed Nord Pool. Later (1996) Finnish companies gained access to it and the Finnish power exchange merged with Nord Pool in 1998. Denmark became a part of Nord Pool in 1999 (Jutland) and 2000 (Zealand). Today power for all four countries is traded on Nord Pool.[1]

It is currently the third largest power exchange market in Europe after Germany (EXX) and France (Powernext). The system price (spot price), which is fixed for every hour, every day at Nord Pool, is only one of three elements in the end-user price for power (the others are net tariff and taxes). Nord Pool is an exchange for professional customers (a marketplace for the wholesale trade in power). In 2004, the value of traded and cleared volume through Nord Pool was approximately €50 billion, which corresponded to almost 2 GWh. Today, the common Nordic power market is the most liquid and transparent in the world. An indication of its advanced state is that end-user prices in Scandinavia reflect spot price development.

Characteristics of the Nordic power market

In order to understand the interaction between Scandinavia and the rest of Europe, it is important to appreciate three characteristics:

1. The power system in Scandinavia is *energy based* with a lot of capacity, but volatile supply. This results in a flat intraday price profile, but large fluctuations in seasonal and annual prices when supply varies.
2. The power system on the Continent is *capacity based*, i.e. new capacity has been developed mainly to cover immediate demand in peak periods. There is always lots of cheap thermal power available in off-peak periods (nights/weekends) resulting in large spot price fluctuations in the course of a day or a week. Short-term marginal costs of coal power determine night and weekend prices, whereas gas power typically determines day prices from Monday to Friday.

   Since the continental system is capacity based, there is also less fluctuation in annual and seasonal prices. Further, there is almost always plenty of capacity available for export to Scandinavia in off-peak periods as long as Scandinavian prices are higher than short-term marginal costs in continental base-load power stations. In the same way, there is always a market for Scandinavian exports to the Continent in peak periods as long as Scandinavian prices can compete with price-determining continental peak-load power stations.
3. Over time, the power exchange between Scandinavia and the Continent will typically involve considerable fluctuation from one year to another as a result of variations in fuel prices, supply and temperature conditions in Scandinavia, and investment cycles in the respective markets.

The price mechanism in the Nord Pool wholesale market, illustrated in Figure 14.1, sorts the production capacity in relation to rising (short-term) marginal costs.

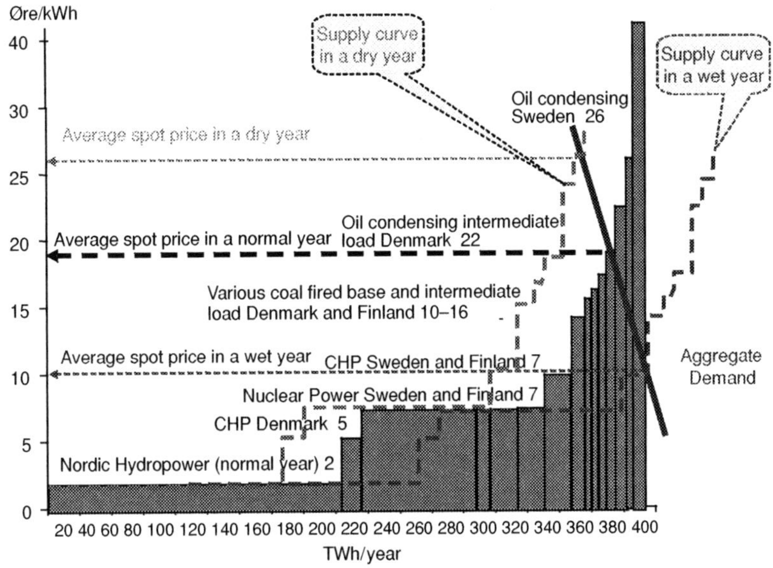

Source: Statkraft

Figure 14.1 Price Formation in Nord Pool

In a year of normal supply and "normal coal prices," total hydroelectric power production will amount to approximately 200 TWh. The combined Scandinavian demand of 380 TWh will result in equilibrium prices of approximately NOK 0.19 per kilowatt-hour (kWh) (or €0.24/kWh). Then the market will typically be balanced with various coal powered base and intermediate capacity.

However, due to the large fluctuations in supply, production may vary by ± 40 TWh per annum. This corresponds to a potential consumer flexibility of 12,000 MW, of which 5,000 MW is in Norway.

In a wet year of high hydroelectric power production, Swedish nuclear power at low marginal costs will determine prices for much of the time. This happened in 2000 when the average annual spot price amounted to approximately NOK 0.1/kWh (€0.13).

In a dry year of low hydroelectric power production, the market has to be balanced with thermal power at high marginal costs, such as gas turbines and oil condensing in Sweden, Finland and Denmark. This was the situation, for instance, in 1996, when it was extremely dry, and the resulting low supply led to an average annual spot price of about NOK 0.26/kWh (€0.33).

World market coal prices may fluctuate considerably, which in turn impacts the marginal costs of coal power. Hence, the market price of coal is of great importance to power prices in Scandinavia. Spot price levels are determined primarily by the supply of precipitation to the hydroelectric power systems in Norway and Sweden. However, increased consumption and exceptionally high coal prices mean that today's power prices are typically higher than before for the otherwise identical supply.

It is extremely important that there is a certain consumer flexibility in a hydroelectric power based system, and that it is accepted that the price mechanism can serve as an instrument of rationing in a difficult situation. The latter is certainly not obvious to the public in general, and in a dry year Statkraft and other power producers are constantly under pressure from politicians and the media. Fortunately, Statkraft's owner (the State of Norway) has acted quite firmly in such situations and thus avoided any populist moves that could damage consumer flexibility.

Power generation investments since liberalization

All in all the deregulation of the Scandinavian power market appears to have worked well. There is a reliable spot price across the Nord Pool area, a dynamic and well-functioning wholesale market, and fierce competition in the end-user market. European players have entered the market and there has been gradual restructuring of the power sector. The question is then why so little has been invested in new capacity since deregulation.

At the beginning of the 1990s there was significant surplus capacity in the Scandinavian electricity system. In anticipation of continuing strong growth in consumption, considerable investment decisions had been made in the 1980s, but the actual growth proved to be smaller.

The fall in prices, which has given the reforms general credibility, is primarily explained by the surplus capacity in the old systems. Surplus is now eroding fast as it is difficult to avoid a growth in electricity consumption in a society increasingly dependent on electrical appliances.

According to Nordel figures for energy for 2004,[2]

- production has increased by 6.2% since 1992
- gross consumption has increased by 10.8%
- average price for the period has been NOK 153/MWh (19.6 euro ¢/MWh)

And for capacity:

- capacity has increased by 7.4% since 1992
- consumption in peak periods has increased by 17.2%

Figure 14.2 presents the average annual growth in generation capacity in Norway from 1960 to 2000. It is interesting to observe that the increases in capacity have consistently decreased from just below 3.5 TWh in the 1960s to 0.25 TWh in the 1990s. Since deregulation, very modest investments in new capacity have been made.

There are no doubt several explanations, but the most important are the low profit margins of new investments and the pressure of various environmental-political obstacles.

The price level for the past decade has been about NOK 0.15/kWh (19 euro ¢/MWh); so historical power prices have not signaled any need to invest in new capacity. Low forward power prices combined with high gas prices have also made it difficult to expect break-even investments in gas power.

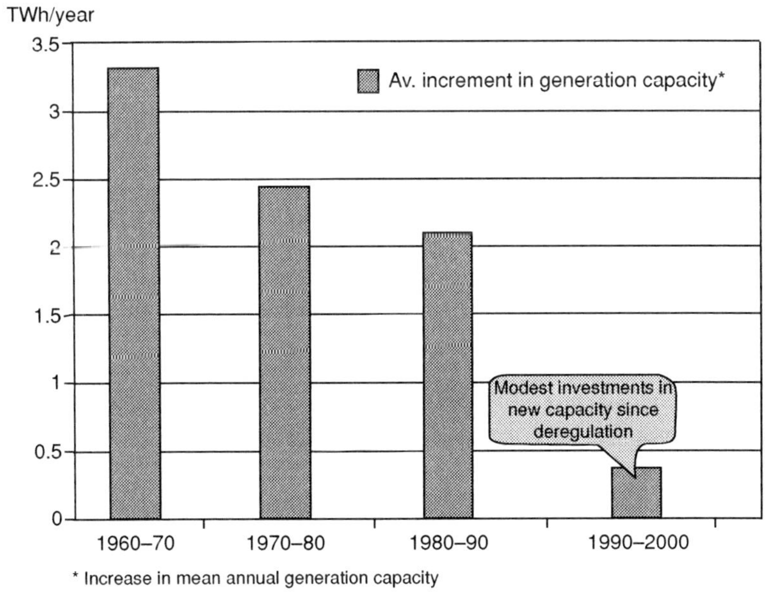

Source: Statkraft

Figure 14.2 Average Annual Growth in Generation Capacity in Norway (1960–2000)

The Norwegian Water Resources and Electricity Board (NVE) knows that new capacity has been planned and that licenses have been applied for, but also that many energy companies are waiting for higher prices and – possibly – other incentives before they will be willing to invest. For instance, there is considerable interest in harnessing wind power in Norway. So far, NVE has licensed 713 MW (2 TWh/year), whereas only 275 MW has been developed and Statkraft is behind 205 MW of them.

According to Econ[3] (2002), there is growing concern over the ability of market-based power systems to generate sufficient timely investments. This is related to the question of whether competitive electricity markets are synonymous with cyclical investment and price patterns. Price shocks in California and the UK market have made it evident that a reliable, stable and efficient electricity market is difficult to put in place.

Trends and challenges in today's investment climate

This section is divided in two and presents the investment climates in Norway and the rest of Scandinavia, followed by those in continental Europe.

Norway and Scandinavia

Statkraft believes that Norway will become a net importer of electricity in a normal year. This is due to a corollary of the combination of market-based power exchange and market-based investments in the power sector.

Because there is almost always a lot of cheap energy available for import in off-peak periods prices in Norway are relatively low, a fact that affects investment decisions. The market-based development of power in Norway, combined with a lot of cheap imports, therefore suggests that Norway will continue to see few investments, and will become a net importer in a normal year.

If investments require a normal rate of return, Norway will probably see an average price level that more or less corresponds to the development costs of new gas power. However, large annual fluctuations in supply will lead to considerable variations in power prices and continued exchange with the Continent – as before.

But in years of normal supply to Scandinavian water reservoirs, a typical week will have approximately 108 hours of continental off-peak prices that will favor imports. Correspondingly, 60 hours of peak period prices will encourage Scandinavian exports to the Continent. In the course of a week, therefore, Norway will typically become a net importer since there are more hours of off-peak than of peak consumption.

An optimally designed Scandinavian market would allow for significant savings in total capacity compared to stand alone systems. Norway, the most hydro-dependent country, used to have a production to consumption ratio of about 105:100 in a normal year. Now, that figure has dropped to around 95:100. Deregulation has therefore delayed investment in new capacity in Norway and Sweden.

Europe

While Cambridge Energy Research Associates (CERA) question the economics of capacity expansion, and market participants constantly call for policy guidance in determining long-term fuel commitments, CCGT/CHP natural gas projects have clearly been the dominant form of capacity expansion since deregulation in the early 1990s. Despite the unprecedented rise in gas prices, CCGT technology is the most economical new entrant technology for mid-merit operation. The high prices mean that CCGTs do not necessarily run base load. In short, the role of natural gas as the fuel of choice for base-load power generation is increasingly called into question because of concerns over price volatility.

Nuclear appears to be the ultimate insurance fuel, with a predictable base-load output and a high share of fixed costs. However, there are widely differing attitudes to nuclear power across Europe, with some countries being for and some utterly against. According to CERA, in the long run nuclear power is poised to enjoy a revival in several European countries, in the wake of the Finnish and French decisions to go ahead with a European pressurized reactor (EPR) unit. In Great Britain, Prime Minister Tony Blair recently announced a revival of the country's nuclear ambitions.

However, a number of issues still need to be addressed before nuclear can become established as a reliable source of power: market risks, high capital costs, uncertainty over financing, doubt about public acceptance, decommissioning costs, the relative inflexibility of nuclear plants and waste disposal issues, for which no solution has been found. In addition, the development and construction time for a new nuclear power station is anything from 8 to 15 years. This means that no significant addition of nuclear capacity can be expected before the middle of the next decade. As a result, in countries where nuclear is not acceptable, coal is valued as an insurance option.

Launched at the beginning of 2005, the European Emissions Trading Scheme (ETS) imposes a cap on CO_2 emissions from approximately 12,000 installations in 2005–07 (first phase), with a view to rolling over and possibly tightening this cap in 2008–12 (second phase). This corresponds to the first budget period of the Kyoto Protocol, which aims to reduce emissions of greenhouse gases in the period 2008–12 to the level of 1990.

The ETS has established a market price for CO_2 emissions. Those companies that have high costs for reducing emissions can demand CO_2 quotas in the market, while those that reduce their production to lower CO_2 emissions can sell their quotas.

CCGT gas power plants have the lowest CO_2 emissions among thermal plants. An increase in production of gas power combined with reduction in production of coal based power is one way of reducing CO_2 emissions in the power sector. However, in 2005 the price of gas increased substantially compared to the price of coal, reducing the former's competitiveness.

At the launch of the ETS, prices stood at about €8/ton of CO_2; they fell briefly to €5/ton, then rose again to peak at €30/ton of CO_2. Since then, prices have fallen, but continue to be more than €20/ton. This volatility is an indication of the great uncertainty prevailing in the market. There are even analyses that estimate a "proper" price level for CO_2 of €50–60/ton. Today's price formation is extremely complex, as it accounts for both the fundamentals of supply and demand and for psychological factors. The market for CO_2 is in an early phase and it is unclear what will happen in the future.

New external conditions, including the use of green certificates and CO_2 quotas, create more uncertainty about the wisdom of investing in new capacity.

Seen in isolation, high CO_2 costs will result in higher thermal marginal costs and power prices in the Scandinavian market, whereas green certificates may well result in a lot of new power at low marginal costs and lower prices.

Discussions are ongoing between Norway and Sweden to establish a Nordic green certificate market. In Norway, the current 10% capital grant incentive to support wind development has not been very successful at promoting plant construction in a country with excellent wind resources. Although definition of the basic market rules is far from final, Norway aims to join the Swedish green certificate system. Creating one market for green certificates would increase liquidity, removing a barrier to the development of the Swedish market, which has been operating since May 2003.

Will renewable energy's prospects seem more attractive in the current environment of very high fossil fuel and power prices? The turmoil in power prices certainly makes renewables seem more attractive for the governments that support them, particularly as their costs to the taxpayer decline. Parallel to the increase in energy prices, the costs of new renewable power continue to fall, and development pipelines to swell. However, over the past 12 months the price of wind energy has increased by 10–20% due to the high demand for wind turbines combined with a lack of manufacturing capacity in the market. As a result, it is expected that for the next three years, the wind market will be a seller's one. The potential for harmonization of support systems is good news for the renewable industry.

The combined effects of volatile power prices, green certificates and a CO_2 market introduced in 2005 will lead to more complex analyses and an increased risk of making investment decisions. Sustainability of green incentives is an issue in itself. The lack (and variety) of green incentives in Norway means that the huge wind potential that exists all along Europe's windiest coastline has so far remained largely unexploited. For these reasons, many investors are watching the market and waiting for more favorable external conditions before implementing new projects.

Statkraft's view on future investments in power generation

High and volatile fossil fuel prices have wide-ranging implications for the power sector. In the short run, they translate into higher prices. The magnitude of the increase varies across Europe, since it depends on the amount of time each power generation technology spends on the margin. Relative price changes also influence power plant dispatch economics and load factors. Generators with a high proportion of fixed-cost technology (nuclear and hydropower) enjoy higher operational margins, as fossil fuel prices push power prices to higher levels.[4]

The longer run implications of the high dependency on fossil fuel are:

- Higher gas prices translate into a higher cost for new-build combined-cycle gas turbines (CCGTs), a technology still considered the least expensive for new capacity investment in many countries.
- The cost of policies to support renewables appears cheaper: higher power prices mean that renewables technologies need reduced complementary revenues in addition to revenue from the power market in order to cover all their costs.
- Because gas prices have increased more than coal prices, it has become more costly to alter the merit order between coal- and gas-fired power plants. Hence, the cost of CO_2 emissions reductions is now higher for the same reduction target.

According to Econ[5] (2005), for the period from January 2005 to January 2008, Nordic production capacity as a whole will increase by about 2 TWh in a normal year. This assumes that in Norway the increase in capacity will be almost 50% hydro and 50% wind for a total of 2 TWh (note that Statkraft's 430 MW Kårstø CCGT project is not included because it is expected to be operational after this period). In Sweden, the nuclear power plant Barsebäck 2 will be closed and the main additions will come from fossil, biomass and wind. In Finland, the additional capacity will be from bio and wind, and in Denmark, primarily from wind.

Statkraft's views on the preferred generating technologies for capacity expansion in Europe are presented in Figure 14.3. In Scandinavia, expansion will primarily be through renewables and green certificates in the short term, with gas coming into the picture a bit farther into the future. In France and Finland, the source of expansion will be primarily nuclear; in Iberia, a combination of wind and gas, in the UK, primarily gas and nuclear, and to some extent wind and renewables. In Poland and the Czech Republic, expansion will come from extending the life of existing coal plants and from new coal projects; in the rest of Europe,

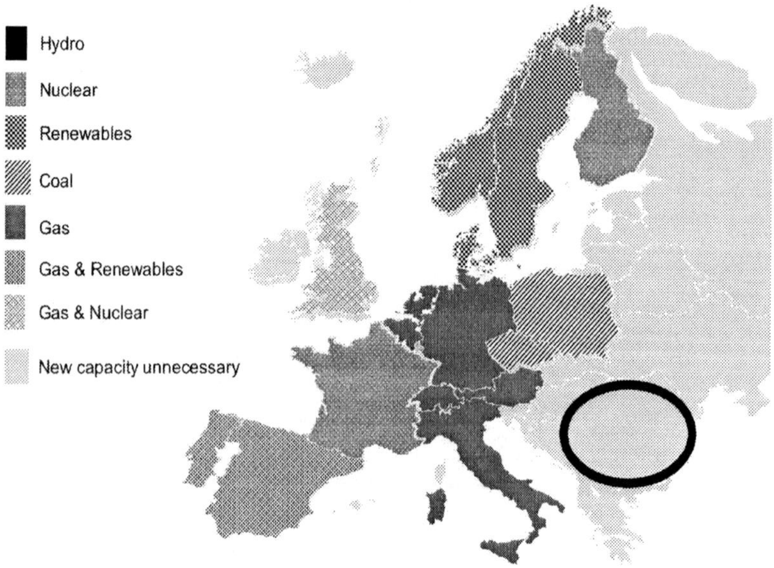

Source: Statkraft

Figure 14.3 Preferred Technology for Expected Capacity Expansion

primarily from natural gas and to a minor extent renewables. Finally, in the Balkans region there is good potential for hydropower.

Statkraft is focusing on the development of new capacity based primarily on three technologies: natural gas CCGTs, wind and hydropower. It would not consider making investments in oil, coal, or nuclear generation. It believes in the importance of asset backed trading and is currently in the process of constructing three CCGT projects, two in Germany (Herdecke and Knapsack) and one in Norway (Kårstø). The company is seeking to develop these gas projects to add flexibility to its current trading offices in Amsterdam and Düsseldorf. It will seek to optimize the dispatch strategy of these new assets and to gain more information in non-transparent markets.

When it comes to renewables, Statkraft has a development portfolio of wind projects in the UK and is seeking new business opportunities for the development of hydropower in South East Europe. In addition, it is beginning to look for opportunities in wind power throughout Europe.

Conclusion

In the 1990s the Nordic electric market was liberalized and is today the most liquid power market in the world. It has done away with excess capacity. However, as we face the future, this development has left one key question unanswered: will markets build adequate and timely capacity? It is evident that the Scandinavian market will require new investments if a price and supply crisis is to be avoided. The investment challenge may well be in different regions – in some places the constraints are on energy, in others on peak capacity. In today's new investment environment for power generation, electric companies must consider the effects of high fuel prices, Kyoto objectives, emission trading schemes and renewable targets.

Gas remains the most economic technology for mid-merit operation, but it faces strong competition from nuclear power for base-load utilization. Nuclear seems poised to enjoy a revival. Coal is an outsider that can play a diversification role in markets where new nuclear is not a viable option. Renewables, particularly wind, are increasingly attractive to policymakers, as the support cost declines under the dual effects of higher power prices and lower development costs owing to technological diffusion, although their development is constrained by resource availability.

In the future the combined effects of volatile power prices, renewable green certificates and the CO_2 market will lead to more complex analyses and an increased risk of making investment decisions. Statkraft expects to play a significant role in the development of new generation capacity investments in CCGTs, wind and hydro, as it seeks to fulfill its vision of becoming a European leader in development of environmentally-friendly energy.

Notes

1. Nord Pool, *Annual Report*, 2004, <www.nordpool.com>.
2. Nordel, *Annual Statistics 2004*, <www.nordel.org>.
3. Econ, *Testing Times: The Future of the Scandinavian Electricity Industry*, Econ Centre for Economic Analysis, Oslo, 2004.
4. CERA, *Fueling the Next Wave of Investment*, Cambridge Energy Research Associates, 2002, <www.cera.com>.
5. Econ, *Nordic Capacity Development 2005–2007*, Econ Centre for Economic Analysis, Oslo, 2005.

15
Emissions Trading – Challenge for Utilities

Stefan Bockamp and Jörg Kruhl

The European Emissions Trading Scheme (ETS) was launched on January 1, 2005. After some initial weaknesses, the new system appears to be operating well and the outlook is positive. This chapter will look at the different aspects of the scheme and the resulting challenges faced by utility companies.

Introduction

The ETS is the first transnational scheme of its kind in the world, and covers the member states of the enlarged European Union. Currently participating in this scheme are more than 12,000 energy-generating and energy-consuming plants, which together account for almost 50% of the EU's total CO_2 emissions.

The aim of the scheme is to achieve the most cost efficient means of compliance with EU commitments under the Kyoto Protocol by using market-based mechanisms and letting the participating companies trade their emission allowances. The EU member states set limits on CO_2 emissions by issuing limited allowances which determine the amount of CO_2 that these companies are allowed to emit. Achieving reduction targets means that a company can sell the "unused" allowances to companies that exceed their limits, or for which emissions reduction measures are too expensive compared with the cost of buying allowances. If a company cannot meet its compliance targets it will be fined €40 to €100 per ton of CO_2 in addition to the missing allowances, which must then also be bought.

Emissions trading – a complex system that is still fraught with unknowns

By the beginning of 2006, emissions trading was up and running, and, apart from a few teething problems, the new system appeared to be working well. The implementation process in Germany could therefore be tentatively described as positive.

In addition to the actual trading of emission allowances,[1] the key to making the scheme a success lies in having a well-balanced allocation process for those allowances. However, Germany's allocation legislation for the trading period from 2005 to 2007 (ZuG 2007[2]) was the outcome of a politically and ideologically driven negotiation process, heavily influenced by lobby groups with widely divergent interests.

As a result, ZuG 2007 is wide open to interpretation. In Germany alone, about 56 different interpretations of the rules were being followed by the end of 2005. Additionally, hundreds of objections proceedings to do with allocation decisions were still pending. At the same time, the German Emissions Trading Agency (Deutsche Emissionshandelsstelle – DEHSt) has been slow to clarify areas of uncertainty. Given that the present certificates are valid only until the end of 2007, there is a danger that by the time the courts issue their final rulings on unresolved issues, the certificates will have lost their value.

The energy sector must therefore contend with high levels of uncertainty. A case in point is the yet-to-be-resolved issue of ex-post adjustments to allocations. It remains unclear whether companies will have to surrender allowances that are surplus to their requirements. As the EU Court of Justice has to decide whether German ex-post rules correspond to the EU Emissions Trading Directive, Germany has decided not to apply any ex-post mechanism until the court's decision is made. Also uncertain is whether the "surplus" allowances that plants surrender – while simultaneously facing tougher compliance requirements – will be returned if the current regime is abandoned.

Hence, the allocation legislation for the second trading period (2008–12) must be simpler to implement, more clearly formulated and have more transparent rules. In particular, the following changes are recommended for this period:

- Germany's Allocation Law (*Zuteilungsgesetz*) should be simplified, avoiding special rules and provisions where possible.
- The base-period data for allocation should be as up-to-date as possible, and fixed nearer to the allocation period.[3]

- To avoid long drawn-out objection proceedings, the allocation process should be more transparent and predictable.
- Given that they form the basis for making long-term investment decisions, the allocation rules must be stable and should not be revoked at a later stage.
- The ramifications of long-term provisions, such as the 14-year exemption from emissions reduction requirements applying to new energy plants, should be reviewed once more.
- Chemical process-related emissions, such as from steel or lime production, currently enjoy an exemption from the legislation. This preferential treatment is arbitrary, lacks transparency and should be re-examined.[4]
- A further element of uncertainty and arbitrariness comes from withholding, possibly for political reasons, reserves that are not released back into the market. Clearer rules and a realistic amount of reserves are required in this area.[5]

With Germany's allocation legislation for the second trading period, ZuG 2012, not due until June 2006, the country's legislators have time to remedy some of these shortcomings.

Emissions trading – implementation and practical experiences

Apart from the volume of allowances allocated to them (see Figure 15.1[6]), emissions trading also impacts the daily operations of energy utilities across many parts of their value chain: the choice of fuel when making short-term power station dispatch decisions; long-term decision-making on new plant construction; monitoring ongoing CO_2 emissions in power plants and making good shortfalls in emissions allowances (energy utilities are "short" as a rule) via the certificates trading market.

In many of these decisions the current, or anticipated, price of emissions certificates plays a decisive role. Before the trading scheme was launched, government policymakers and business leaders alike expected liquidity in the market to increase gradually off the back of moderate prices.

However, it has since become clear that high liquidity is not a realistic expectation because the market is, by definition, "short." There is a shortfall in supply, which will persist as long as there are no short-term options for reducing emissions. This has probably been further exacerbated by the cuts in allocation plans for the countries of Eastern Europe (see Figure 15.2). That is also why the first half of 2005 was characterized by strong demand from companies anticipating a significant

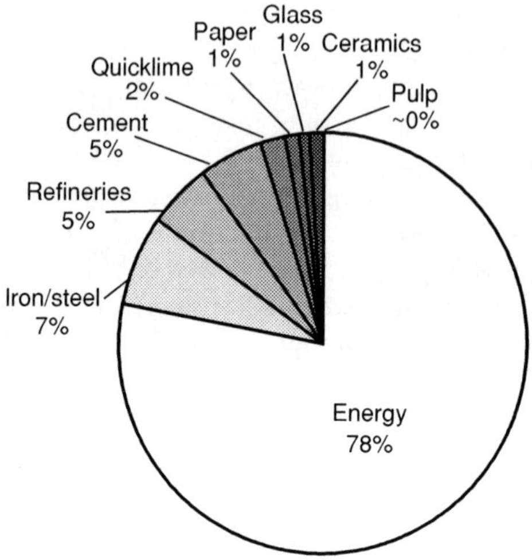

Source: Adapted from DEHSt

Figure 15.1 Impact on Energy Producers

short position for the period to 2007, and wanting to address this shortfall as soon as possible. But supply was non-existent because of the delayed launch of trading registers in several countries, and the absence of any pressure on market players to divest surplus emission allowances in a timely fashion. Europe's actual shortfall of allocated allowances versus required allowances became clearer after the first year of trading and the official EU CO_2 emissions report for the year 2005 in April 2006. Due to low consumption of certificates the young market "saw" a surplus and reacted with falling prices (see Figure 15.2). E.ON Energie has positioned itself on both sides of the market – as a buyer and seller – to learn about the new system.

The prevailing view in the industry is that the only feasible way to compensate for allowance shortfalls of the order of several tens of millions of metric tons of CO_2, is to switch fuels in power stations. Given the relative price of natural gas to coal, this applies particularly to the UK. Fuel prices for coal and natural gas determine the costs of CO_2 reduction measures and hence the price of CO_2 certificates. For example, an energy company will dispatch a gas-fired power station before a coal-fired station only if the sale of the emission allowances freed up by the substitution generates sufficient revenue to cover the extra costs involved. Hence, emissions certificates must be taken into account in power station dispatch decision-making if emissions reductions are to be achieved by switching fuels.

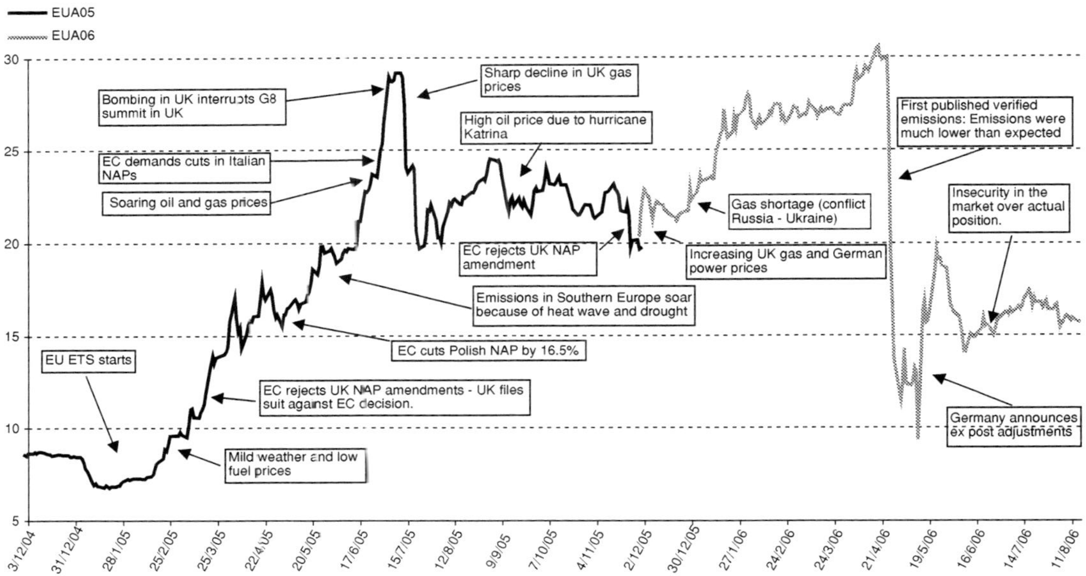

Note: NAP = national allocation plan

Source: E.ON Sales & Trading GmbH, Munich, Department of Analytical Support, April 2006

Figure 15.2 Price History and Major Events in EU Allowance Trading

Emissions trading – monitoring and reporting

In addition to facilitating the trade in certificates (and litigation), the new scheme also has ramifications for the core business of energy utilities – operating power stations. This is because Germany's legislators adopted the European Monitoring and Reporting Guidelines into German law without giving due consideration to local conditions. No adjustments were made for long-established power station operating practices because, when formulating Germany's statutory framework, legislators focused instead on the issue of allocation.

Recommendations by energy-generator industry associations to simplify and adapt the rules to tried and proven practices in Germany fell on deaf ears.[7] Adopting these recommendations might have streamlined the monitoring processes without reducing their accuracy to any significant degree. Much could be achieved were the system to recognize pre-existing, tested data from other reporting processes (such as those applicable to co-generation plants) and dispense with accreditation requirements for testing laboratories. Germany's state authorities and energy companies are currently in the midst of a learning process, made more difficult because responsibility for the approval of monitoring plans is divided between state and federal authorities.

Emissions trading – current market development

The Carbon Market will take time to develop. Price volatility is extremely high and heavily dependent on political frameworks, changing fuel price differentials and weather conditions. New players, like banks and hedge funds, have already entered the market, bringing in speculation as well as their business approach.

Today power plant operators are required to maintain a detailed CO_2-balance monitoring system, starting with fuel purchasing and use and covering all stages of the production chain. The system must also stand up to scrutiny by the certifying body. In addition, each power station must keep a greenhouse gas emissions balance sheet.

The utility industry has always faced the difficult, but vital, choice of either generating electricity in its own plants or purchasing power on the open market. The complexity of this make-or-buy decision has now been considerably increased by an extra factor: CO_2 certificate prices (see Figure 15.3), which will make it more difficult to assess commercial viability as prices follow different market laws, yet also influence daily power station merit-order planning.

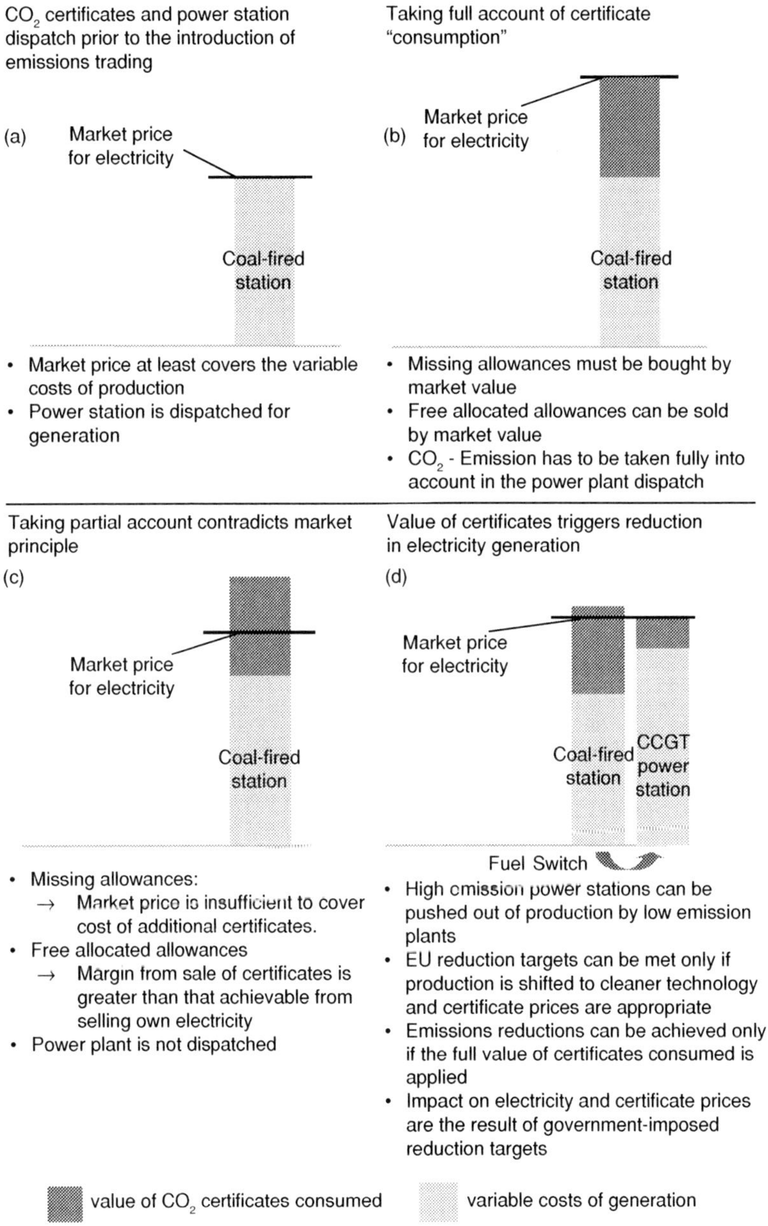

Source: E.ON Energie AG

Figure 15.3 The Make or Buy Decision

In recent months, there has been a heated debate in the industry sector as to whether, and to what extent, certificate prices affect electricity prices. Germany's Federal Cartel Office is also now reviewing the issue, investigating allegations that German energy companies have a dominant position and employ anti-competitive practices to manipulate markets and prices.

Emissions trading needs emissions reduction – a cap and trade system

Options for acquiring emissions allowances are extremely limited (see Figure 15.4). Apart from buying on the open market, the only alternative is investing in emission reduction funds or running international emission-reduction programs under the Joint Implementation (JI[8]) and Clean Development Mechanism (CDM[9]) frameworks.

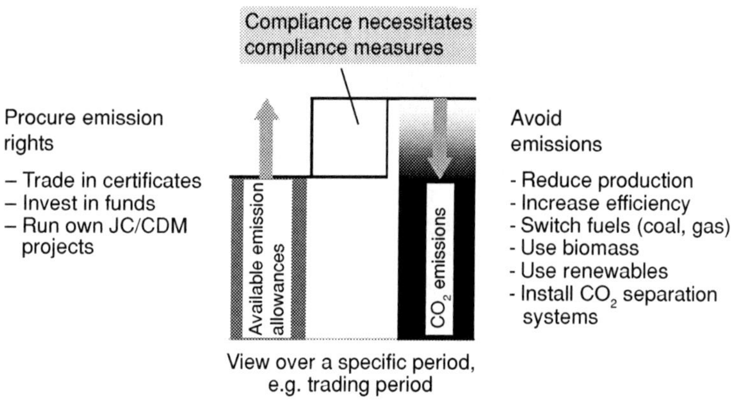

Source: E.ON Energie AG

Figure 15.4 Compliance Options for Electric Utilities

JI and CDM are two of the so-called flexible mechanisms of the Kyoto Protocol designed to allow its parties flexibility in achieving their emission limitation and reduction commitments. Under these mechanisms, projects that reduce emissions or remove carbon dioxide from the atmosphere generate emission certificates: Certified Emission Reductions (CERs) in the case of the CDM and Emission Reduction Units (ERUs) under JI.

If officially recognized, these measures may yield CO_2 credits. But acceptance is not guaranteed and applying for recognition requires a great deal of time and effort. Alternatively, there is the option of physically reducing emissions. Besides throttling back production – obviously not a viable large-scale solution – ways of reducing emissions include fuel efficiency measures such as fuel switching, biomass and long-term programs for building new power plants (possibly with CO_2 separation and sequestration technology).

The difficulty is the very limited range of short-term options. Ultimately, energy companies can only reduce emissions by running their own projects and building new power stations. But here the lengthy implementation time frames involved pose a major problem. There are few projects that can be completed before the end of the first trading period, and their emissions-reduction potential is small. Yet at the same time, the current policy and regulatory environment does not provide the level of certainty needed for large-scale capital investment in effective, long-term measures.

Emissions trading – local responsibility versus global climate protection

Germany's energy sector takes its responsibilities for resource conservation and environmental protection very seriously. This has made it a world leader in efficient electricity and heat production. Although the global warming potential of man-made CO_2 emissions remains uncertain, the "no-regrets" approach requires that action be taken. But a balance must be struck between the three equally important goals of environmental protection, economic viability and a secure energy supply.

The means adopted to achieve the Kyoto targets must however be global, not national, in scope. It would be a major mistake to force Europe's energy sector to become heavily dependent on natural gas as this would expose the European economy to incalculable long-term political risks. Rather, economic measures should be employed primarily in those countries (such as Russia, China and the USA) where power station efficiencies, as shown in Figure 15.5,[10] are much below German levels.

Germany already has the most efficient power stations in the world, and efficiency will automatically increase as utilities implement plans to upgrade their generation fleet. Emissions reduction targets over and above will entail massive additional costs and may force the country to abandon the security of its diversified energy mix.[11]

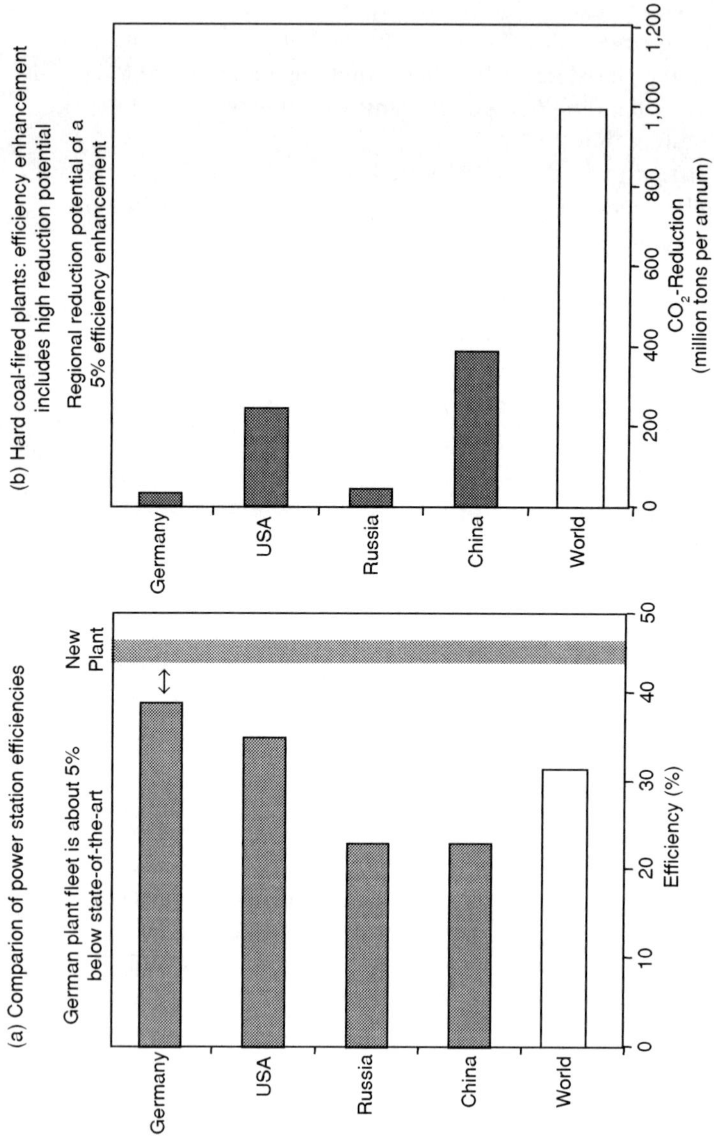

Source: Adapted from Schilling: *EnergieFakten*, IEA Key World Energy Statistics

Figure 15.5 Power Station Efficiencies

In the long run, it will also be necessary to integrate other German CO_2-reduction measures into the emissions trading system. The current mix of economic instruments including emissions trading, the Renewable Energy Law (EEG), Co-generation Law (KWKG) and so on will have to be simplified as well.

That many countries do not participate in the emissions trading system is just one of several factors hampering its effectiveness as an economic model. In the first trading period, the EU's own trading scheme does not cover even 50% of the world's CO_2 emissions. Thus the emissions reduction contribution of the EU transport and domestic household sectors are not the result of free market forces. Rather, it is a negotiated position lying along a continuum between an ideologically driven desire to achieve maximum reductions at one end, and purely economic interests at the other.

The experience of the lead-up to, and implementation of, the emissions trading scheme so far clearly identifies some steps that should be taken if reducing greenhouse gas emissions is to continue to play a global role in the years ahead. The following recommendations apply to both the Kyoto period and the post-Kyoto period (after 2012):

- CO_2-reduction targets should not be set so high as to be economically untenable or to compromise the security of energy supply.
- The risks associated with fuel procurement for nationwide electricity supply in Germany and the rest of Europe should be minimized by, for example, maintaining a diversified energy mix.
- Efficiency gains and the use of low-emission energy sources should be commercially worthwhile (incentive effect).
- Insofar as is possible, Europe's rules should be harmonized with prevailing global rules to prevent competitive distortions.
- The scope of the system should include the emissions reduction capacity of all industry sectors.
- Measures should be taken to harness the existing worldwide emissions reduction potential. This might be achieved through a global trading system, or by simplifying the means of implementing international projects.

Emissions trading – including all global potentials

Of course, one must also look beyond Germany's national borders. The Federal Republic, with a generation fleet on average only five percentage points below the current state-of-the-art for power plant technology, is not representative of the global situation.

In future, the pending modernization of power stations is likely to free up major potential for emission reductions. Over the next 15 to 20 years, Germany needs to replace at least 20,000 MW of fossil fuel-fired generation capacity. This represents a significant opportunity to achieve substantial efficiency gains. Assuming that: (1) lignite continues to be used even though it produces the highest specific emissions, and (2) only about a third of current hard coal use is replaced by gas, which seems fairly realistic, then as shown in Figure 15.6, in the energy supply industry alone we can expect an additional 10% reduction in CO_2 emissions from improved power plant efficiency up to 2030.

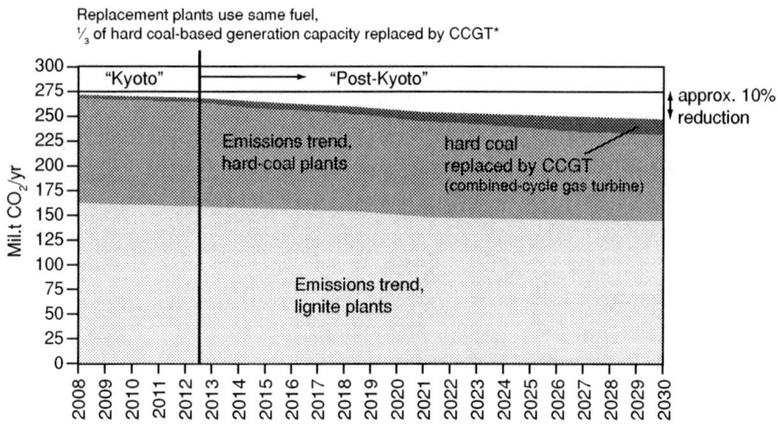

* Plant replacement after 45 years, old plant utilization rate: Lignite 7,500 h/yr, HC 4,500 h/yr
Efficiency: old plants as per actual plant efficiency; new plants – Lig. 43%, HC 45%, CCGT 58.5% in 2010, Progress through innovation: 0.25% pts/year (Lignite, HC); 0.125% pts/yr for CCGT

Source: E.ON Energie AG

Figure 15.6 Efficiency Gains Automatically Result in CO_2 Reduction

Emissions levels and the scope for emissions reductions vary considerably from region to region. The global nature of the CO_2 problem, and the lessons emerging from Europe's new emissions trading system, show that the international community must unquestionably focus its reduction efforts on those areas that yield the greatest benefit for the least cost. If the entire world were to take the same measures that Germany will implement in the next 15 years, this alone would yield a global reduction in CO_2 emissions of about 1 billion metric tons per year.

Conclusion

The new emissions trading system has been in place since January 1, 2005. By the beginning of 2006, the practical experience of application and allocation processes for emissions rights showed that unless the system was made much simpler, it would not yield the benefits sought through emissions trading, i.e. the reduction of emissions in an economic and efficient way by introducing a market for CO_2 certificates. The allocation rules must be made clearer, simpler, and more transparent.

All participants were surprised by the price trends in the market for CO_2 emission rights. And yet the reasons are clear. The substantial cuts made to planned allocations by politicians led to a higher "short position" than anticipated, while the lack of available trading registers kept some countries with a surplus of certificates out of the market. Finally, rising fuel prices made the reduction measures required across Europe more expensive.

Germany's energy sector is fully committed to meeting its CO_2 reduction targets because doing so promises the triple benefit of economic viability, security of supply and environmental protection. However, further increases in these targets would effectively compromise the diversity of Germany's energy mix, resulting in a greater dependence on imports, increased exposure to price fluctuations, and consequently a higher potential for political risk.

It is important that any initiatives to change reduction targets give due consideration to the economic ramifications. Excessive reduction commitments can lead to disproportionately high costs, thus posing a huge risk to energy supply. Accordingly, international initiatives – to increase the efficiency of electricity generation in other parts of the world, for instance – should be part of a global cost-minimization strategy.

Notes

1. An emissions allowance refers to an allowance to emit one ton of carbon dioxide equivalent during a specified period (2005–07 or 2008–12), which shall be valid only for the purposes of meeting the requirements of the EU Emissions Trading Directive and shall be transferable and tradable in accordance with the provisions of the Emissions Trading Directive.
2. ZuG2007 is the German law on the national allocation plan for the period 2005–07. It defines the rules to be applied in allocating allowances to plant operators working under the scope of the EU Emissions Trading Scheme.
3. The original base period used by the current legislation was 2000–02. The corresponding trading period is 2005–07. If the original base period is applied

for the next trading period (2008–12), the base-period data will lag the beginning of the trading period by six years.

4. The strength of emissions trading is that it is market based and will lead to the most efficient reduction measures. If some industries are free of reduction targets, thus shifting the burden to other sectors, this will be compromised.
5. Reserves are used to provide free allocations to newcomers. If the reserve is too large, and not all allowances are released to the market, scarcity will increase leading to higher prices as reduction targets remain unchanged.
6. DEHSt, *Emissions Trading in Germany: Allocation of Allowances for the First Commitment Period 2005–2007*, Deutsche Emissionshandelsstelle, Berlin, February 28, 2005.
7. VGB, *Determination of CO_2 Emissions of Combustion Installations (CO_2 Monitoring)*, VGB PowerTech, Essen, August 27, 2003 (to be ordered at VGB-agency; text only in German).
8. Joint Implementation under Article 6 of the Kyoto Protocol allows an entity (government or authorized company) in a developed country to invest in emissions-reducing projects in another developed country. The emission reductions can be used by the purchaser to meet its targets.
9. The Clean Development Mechanism under Article 12 of the Kyoto Protocol provides incentives for clean investment in non-Annex I Parties (mostly developing countries) in return for credits ("CERs" – Certified Emission Reductions) that can be used by Annex I Parties (developed countries) towards meeting their Kyoto targets.
10. Data for China, Russia, World and Germany is taken from H.D. Schilling, *Wie Haben Sich Die Wirkungsgrade Der Kohlekraftwerke Entwickelt Und Was Ist Künftig Zu Erwarten?* 2004. See <www.energie-fakten.de> accessed February 20, 2004. Data for the USA is taken from *Wirkungsgrade Von Steinkohlekraftwerken Im Jahr 2004*, in <www.rag.de>, section sustainability, chart 06. Data for new plants is taken from IPTS-EU Commission, *Integrated Pollution Prevention and Control (IPCC) – Reference Document on Best Available Techniques for Large Combustion Plants*, Institute for Prospective Technological Studies, Seville, May 2005.
11. Germany's energy mix in 2005: nuclear (29%); lignite (26%); hard coal (21%); natural gas (10%); renewable energy sources (11%); rest (3%).

Part Four

Beyond the Mere Technical – A Broader Perspective

16

Policy Responses for the Future of Energy: Global Overview and Perspectives

Subhes C. Bhattacharyya

This chapter analyzes the major global challenges facing the energy sector and discusses possible policy responses. Analyzing the oil scarcity issue, the chapter suggests that killing the goose that lays the golden egg may trigger an energy transition away from oil domination but that such a change will bring major, and perhaps violent, global changes. The problem becomes complicated as the policies involving the three Rs of the 1990s (right prices, reform, and renewable energies) fail to yield the desired results. The chapter calls for alternative strategies and policies for providing energy access to the energy have-nots of the world and for better functioning of the energy industry.

Introduction

Investigation into the future of energy has become a fertile area in recent times due to soaring oil prices, which reached unprecedented levels in 2005, and the belief that high oil prices may become a permanent feature of our industry. In a dynamic world where profound political, social, economic, technological and even ideological changes shape our present and future lives and living conditions, changes in the energy scenario are natural and inevitable. But the cost and pain associated with adjustment and the fear of being caught unprepared make research in the area interesting.

Although issues related to security of supply have dominated recent discussions and drawn public attention, they do not constitute the only challenges facing the global energy sector. While "energy haves" are concerned about future prospects, billions of others are struggling to get access to affordable, reliable, and acceptable energy services. Access to energy has been identified as a major challenge if sustainable worldwide development is to be achieved. At the same time, unsustainable practices abound: unprecedented growth in demand arising from new as opposed to traditional centers of growth; the globalization of wasteful consumption patterns; the concentration of supply in politically unstable regions; financially bankrupt state entities perpetuating supply, etc. Governance issues, restructuring and reform of the sector and environmental concerns remain relevant. This complex set of challenges will shape the future of the energy sector and, consequently, the sector should perhaps charter a new path along which its activities will be organized and performed differently.

Possible structural and behavioral changes bring new issues to the surface. Will this herald an end to the scale and scope economies enjoyed by some segments of the sector? Will it end the top-down policies and practices followed so far? Will the state have any role to play in the future? Will the poor remain marginalized or do they stand to gain? Will energy services become affordable, reliable, acceptable and equitable?

The objective of this chapter is to put these challenges in their global context and proffer an overview of possible policy responses. It attempts to provide a panoramic view of energy sector issues by identifying the drivers behind them. The key policy responses are then presented keeping in mind various constraints.

The chapter links the great transitions in energy use of the past to current policy options and potential future transitions with regard to oil scarcity, supply security, energy access, environmental concerns, the use of renewables and the need for reforms.

Energy transition and future challenges

Energy demand patterns have been undergoing significant changes ever since mankind started using energy. Prior to the Industrial Revolution, energy was mostly derived from natural energy flows such as water and wind, and human and animal power, and restricted availability limited the amount of energy that could be used. Energy conversion also had limited scope – conversion from chemical energy to heat and light and chemical energy itself were derived from natural sources – animal or plants. Energy consumption typically did not exceed 0.5 tons of oil equivalent (toe) per capita per year.[1]

Since then, two major transitions have shaped energy systems at all levels. The first involved a shift towards coal, made possible by a radical technological innovation – the steam engine powered by coal. This allowed fossil energy to be converted into work and made the supply of energy site independent, as coal could be transported and stored as needed. The first application of stationary steam engines was to lift water from coal mines, thus improving productivity. Later, they were instrumental in the development of an entirely different form of production organization: the factory system. Mobile steam engines, on locomotives and ships, brought a transport revolution. Railway networks were laid in various countries making the supply of energy and productive resources easier and less site dependent. Ships were converted from sail to steam engines. By the middle of the nineteenth century the steam age had replaced traditional non-fossil energy sources in the industrialized countries. Energy consumption increased as well and was about 2 toe per capita per year in this period.

The second transition was triggered by a number of innovations that resulted in a diversification of supply options and end-use technologies. Electricity was the most important innovation of this period. It allowed the easy conversion of energy to light, heat or work at the point of use. Another crucial invention, the internal combustion engine, revolutionized individual and collective mobility through the use of cars, buses and aircraft. The availability of new technologies allowed new energy supply and petroleum, which was discovered in the mid nineteenth century, found new applications. As technology progressed over the next fifty years or so, oil emerged as the dominant energy form.

In both these transitions, scarcity of energy in the sense of physical exhaustion of resources was not part of the equation. Biomass and wood fuel were and are available in quantities equivalent to if not more than what were used before the Industrial Revolution. Coal provided an alternative to the traditional sources of energy and gained acceptance because of higher energy density, flexibility and mobility. Similarly, the second transition did not happen because of any threat of scarcity of coal. Coal was abundantly available and remains so to this day. But it could not compete with oil in terms of ease of use, versatility and universal appeal. Oil has maintained its position as the dominant energy source for over four decades.

These two great transitions also brought far-reaching structural changes in terms of industrialization and urbanization. Economies started to move away from agriculture towards industry and manufacturing. The present trend is away from basic industries to services and information-based industries. Urbanization has led to profound changes in the world's economies in terms of economic activities, lifestyles, social values, and relocation from rural to urban areas. The transitions also radically altered energy demand and supply. Industrialization, urbanization and increasing monetization of economic activities led to a shift towards commercial energies and away from traditional energies. Demand for flexible, convenient and cleaner energy forms increased. This also resulted in higher per capita energy demand but a decrease in energy intensity.

Yet, the benefits of the transitions did not reach everybody, as about a third of the world's population still relies on traditional energies. The poor, particularly in rural areas, are disadvantaged in this regard and the lack of access to energy imposes undue burdens on children and women. At the same time, uncontrolled urbanization, rapid population growth, changes in economic and social structures, and imitation of developed country lifestyles are imposing additional burdens on energy supply. But politically motivated pricing, revenue-generating tax policies, inefficient functioning of the energy market due to inappropriate market structure or collusive behavior and the poor performance of firms in many areas bring additional problems to the sector. Moreover, restructuring efforts to fix some of the performance-related problems did not progress well in most parts of the world and new issues are arising in places where they did. Activities related to supply and consumption of energy caused environmental damage at local, regional and global levels. While billions of people are deprived of a modern energy supply, 140 bcm of natural gas (equivalent to more than 100 million tons of oil) are flared around the world every year, wasting a valuable resource and damaging the environment.

Any future transition is expected to emerge from the present day issues and challenges. High reliance on oil centralized in a single region has led to concerns about scarcity and security of supply. Unprecedented growth in demand, skewed regional distribution of demand, profound divergence in consumption, 2 billion people without access to clean energy, and multidimensional concerns about environmental damage due to energy use – in short the symptoms of unsustainable energy practices will continue to haunt us. It is obvious that the energy sector requires a paradigm shift if it is to return to a sustainable path. This implies that the sector has to organize itself differently and to change practices and policies at various levels.

Policy responses to future energy challenges

Oil scarcity

From the perspective of energy transition, the pertinent questions about the future of energy are when the next major change can be expected and what will emerge as the next dominant fuel. Recent outpourings of information about the impending scarcity of oil[2] tend to suggest that oil's domination of the global energy market may not last long. The concentration of oil and gas in two regions – the Middle East and Former Soviet Union – the smallest reserve to production ratio (R/P) ever for oil in general, a ten year R/P ratio in North America and the possibility of an even quicker depletion in Europe have fueled recent speculation that global oil supplies are running out. The high concentration of reserves in a single region adds a political dimension to the problem.

Oil scarcity has a short-term and a long-term dimension. The recent high prices may be a short-term phenomenon arising out of lack of investment in exploration, production and refining capacities to replace old and aging facilities and ensure surplus capacity to meet future demand. The low variable costs of oil production and the application of outdated rules encourage an inherent state of overproduction, which brings with it price crashes and consequent losses to investors. Three main factors – restricted access of international oil companies to prime oil acreages since the widespread nationalization of oil industries in the 1970s; OPEC's market coordination policy, leaving the market to non-OPEC producers with expensive oil; and the erosion of excess capacity due to war, conflicts and unprecedented demand growth – have contributed to lower capacity expansion in the past decade. Demand management through price mechanism when capacity is limited leads to price spikes of the kind we have experienced recently but the expectation of such a situation will lead to an increase in capacity, bringing prices down again. Policies to improve investment climate, the removal of restrictions on investments in the main oil bearing areas (including the removal of political barriers on companies wanting to invest), and investment protection arrangements would be required to overcome the short-term problems. Similarly, improving the financial management of national oil companies (NOCs), capacity building and reforming the NOCs might help.

The longer-term problem is a more serious concern. If enough oil is not there in the ground, relying on oil is not a logical strategy. Yet, this is not the first time that concerns about depletion have hit us. One wonders, despite limited oil and gas reserves, how the USA has been able to maintain a fairly steady R/P ratio of ten years since 1980. Oil production there has declined from around 10 million barrels per day to 7.7 million barrels per day while proved reserves have reduced from 36.5 billion barrels to 30 billion barrels. This was made possible by the additional discovery of around 65 billion barrels during this period, without any major reserve finding and technological breakthrough. Resource-rich regions with much less exploratory activities may stand a better chance of similar performance. The hope lies in the fact that both reserve and resource estimates are essentially guesses and conservatism reigns supreme in such areas. Technological changes and economic conditions would improve the reserve estimates, pushing the depletion horizon back further.

Although North America and OECD Europe have maintained a high level of output compared to reserves for more than a decade now,[3] low prospects of major new discoveries in these intensively explored areas and a precariously low R/P ratio (between five and ten years) mean that the depletion of existing resources in these regions is imminent. Consequently, their import dependence will increase significantly from present levels, exposing their economies to the vagaries of the international oil markets. The propaganda about global oil depletion, essentially based on the above regional depletion phenomenon, is an attempt to attract public attention and speculation, and to create panic. The OECD countries' increased dependence on imported oil at a time of growing demand in Asia is making the market nervous.

From a demand perspective, oil domination has perhaps already started to decline. According to the IEA in 2004,[4] oil only had a 35% share in the global primary energy supply (PES) of 10,230 million tons of oil equivalent in 2002, closely followed by coal (23%) and gas (21%). Oil dominates primary energy supply significantly in the Middle East (56% of supply), followed by OECD Pacific region (47%), Latin America (47%), OECD North America (40%) and OECD Europe (38%). Everywhere else, some other fuel is as important as or more important than oil.[5] (See Figure 16.1[6]).

A careful look at the evolution of global energy indicates that the substitution of oil in electricity generation has continued since the oil shocks of the 1970s and that oil has never regained its lost market (Figure 16.2). Domination of oil at the final energy consumption level arises because of the near-total dependence on oil in the transportation sector (see Figure 16.3). Thus the appearance of any viable substitute in transportation could bring the end of the oil era. High oil prices provide the necessary encouragement to seek alternatives. Already hybrid

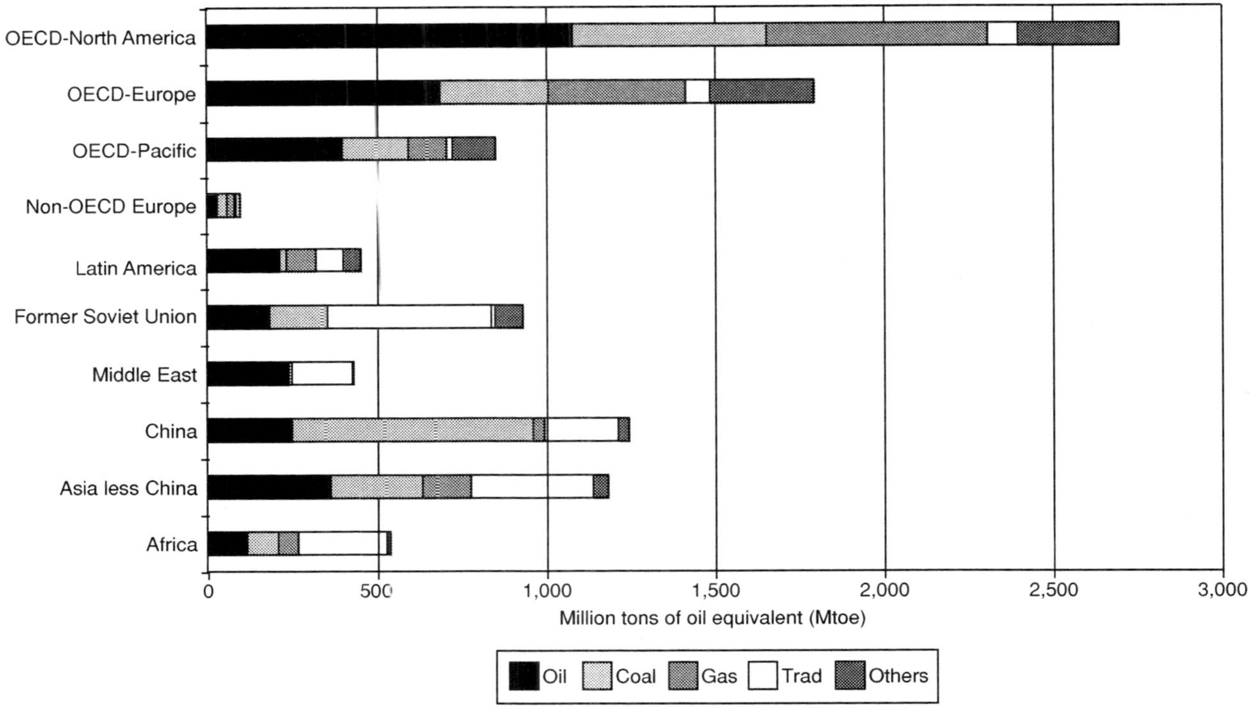

Source: IEA, Key World Energy Statistics, 2004

Figure 16.1 Regional Composition of Primary Energy Supply in 2002

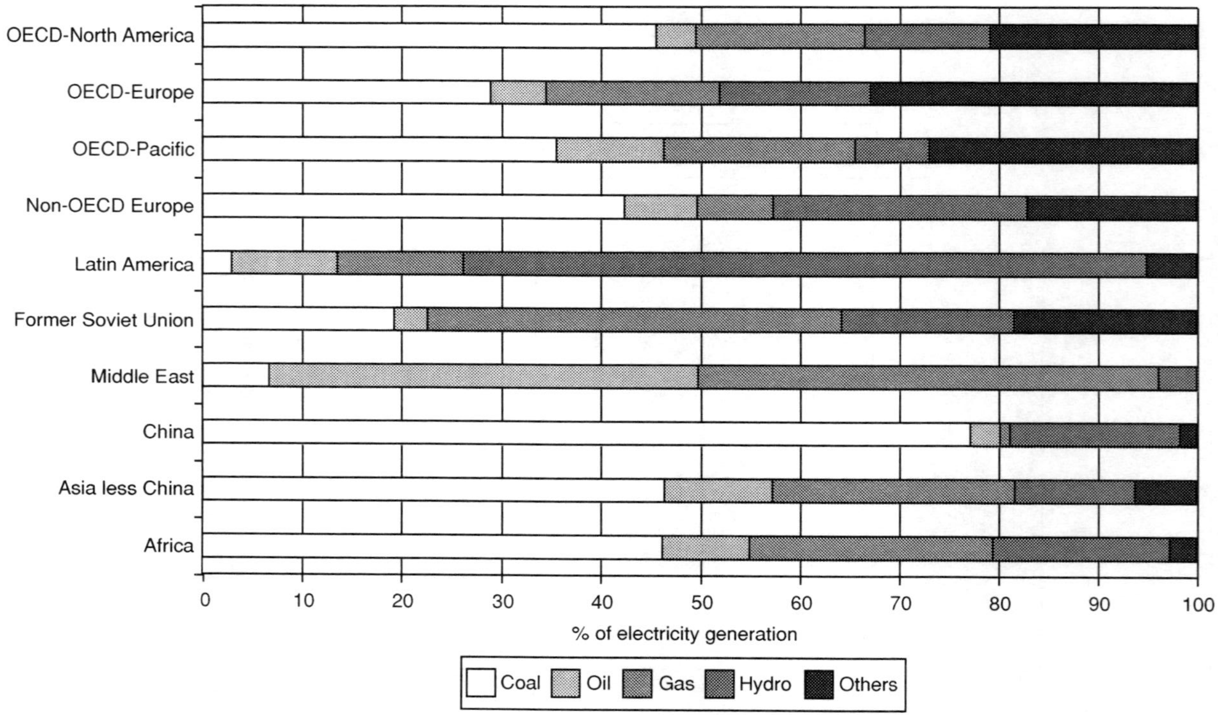

Source: IEA, Key World Energy Statistics, 2004

Figure 16.2 Regional Variation in Electricity Generation Mix

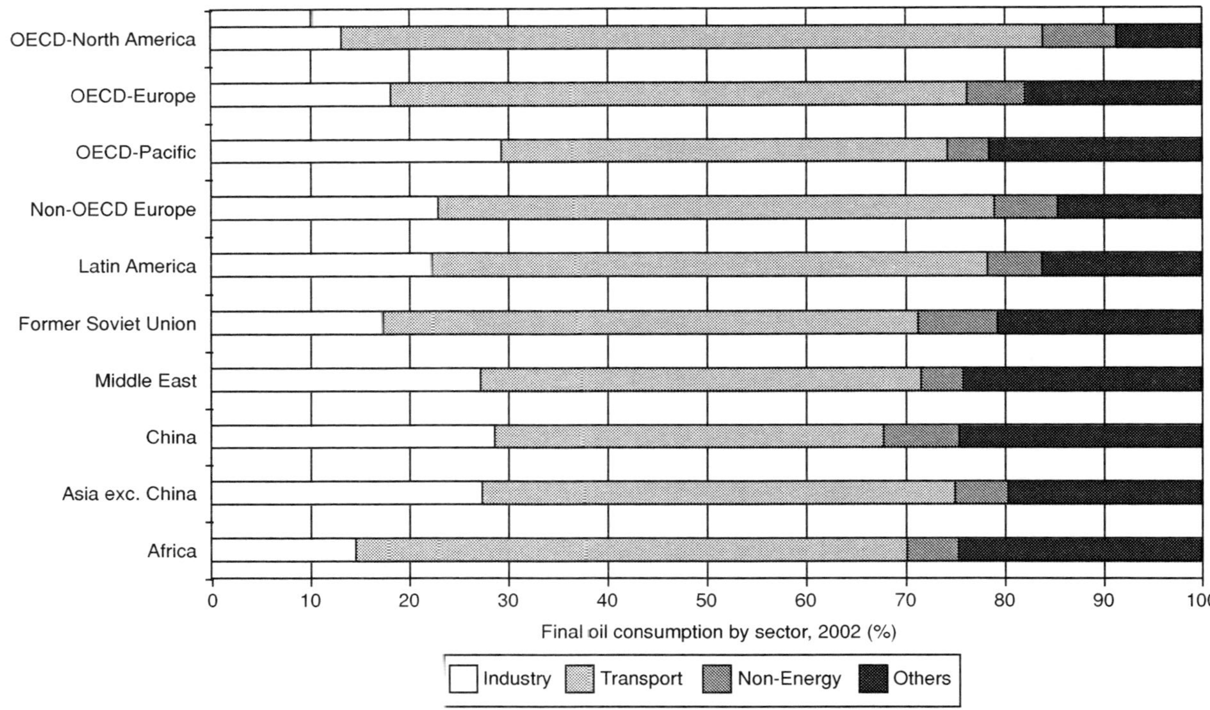

Source: IEA, Key World Energy Statistics, 2004

Figure 16.3 Final Oil Consumption by Sector and Region, 2002

fuels, biofuels and derived fuels are making desperate attempts to break even. And there does not need to be a single universal alternative – a combination of substitutes in separate niche markets would be sufficient to ensure the economic abandonment of oil. This would be in line with the lessons learned from previous transitions.

A viable substitute for oil in the transport sector would bring a sea change in the energy markets: importing countries would be freed from security concerns, while exporting countries would discover that the goose was no longer laying golden eggs. Countries sitting on huge oil reserves and hoping for windfall gains in the future when oil became scarce in other areas would be the worst hit. The transition could be violent and could make the world a much less safe place. However, it is naïve to think that the oil-rich countries would sit back and watch such a disastrous scenario develop. Flooding the market with cheap oil remains their main weapon, which could be unleashed at any time, thereby making costly investments in alternative energies unviable. Although such a strategy is not new in the history of the oil industry, it is doubtful whether low oil prices can be sustained, which makes the driving out of costly alternative sources difficult. Thus a painful end of the era of oil domination is quite likely, though when such a transition might occur remains a matter of futurology.

Policies to ensure supply security

A more immediate concern is related to ensuring adequate energy supply in an affordable and reliable manner to meet future demand. Two major changes can be noticed in the evolution of patterns of demand: a shift towards more reliance on electrical energy by final users and a geographical shift in demand towards the Asian region. These two changes require more capacity addition in the conversion processes and facilities closer to demand centers.

Global energy demand is projected to grow at an annual rate of 1.7% between 2000 and 2030, thereby requiring about 1.67 times 2000 energy levels by 2030.[7] Developing countries' demand would increase to 43% of global primary energy in 2030 as opposed to 30% in 2000. According to IEA (2003),[8] the energy sector will require an investment of $16 trillion between 2001 and 2030, of which about 60% ($10 trillion) would be in the electricity industry, with 19% ($3 trillion) each for oil and gas. About half of this need will come from developing countries. Mobilizing such huge investments in the right place at the right time is a major challenge, particularly in developing countries, which would need to step up their investments many times over. The recent slump in private investment in the energy sector is not an encouraging sign.

The problem is further aggravated by the following:

1. Financial difficulties of the national energy companies, especially the electric utilities, due to poor pricing policies, poor management and political interference in decision-making.
2. Budget constraints on governments, forcing them to commit less funds to energy sector projects, and higher levels of competition for funds due to the refocusing of state activities and social needs.
3. Unfinished reforms of the energy sector in many countries creating a prolonged period of transition, during which the state withdraws from the sector expecting private investment to come, while private investment does not materialize because of the unstable environment. Investments cannot wait until reforms are finished, but any reform depends on political stability and willingness.
4. Heightened security concerns for energy installations in many countries due to political or other conflicts. It is reported that such concerns have cost Saudi Arabia billions of dollars to protect its oil installations in recent years and international oil prices now bear a risk premium of $10–15 per barrel, much higher than the cost of oil production in many areas.

The global phenomenon of the 1990s was the emphasis placed on the three Rs: energy market Reforms, Renewable energy technologies, and Right prices. Reform has not been a great success in many countries and state funding for energy has declined, without any concomitant participation from the private sector. Depending on reforms to solve the energy access problem, therefore, does not appear to be a logical approach. Policies to avert the problem could involve a return of state-led investment in major energy projects, especially those that require longer-term financial commitment and are of strategic importance. As restructuring and reform are failing to deliver in many countries, state presence is becoming more important, especially in developing countries that did not benefit from the inflow of private capital in the energy industry and where private capital did not develop nationally desired energy sources such as hydro, nuclear and the like. Similarly, instead of wasting decades on reforms that can hardly be implemented, it may be better to focus on better management of state agencies through enterprise level changes.

The right pricing for energy and the promotion of renewable energies have been at the forefront of policy for quite some time. Removal of price distortions has proved difficult in developed and developing countries alike. Distortions in the energy markets, distorting taxes for revenue generation, and price distortion for social or other reasons militate against correct price signals. Relying on such a non-attainable goal for enhancing energy security does not make sense. The removal of trade barriers and flooding markets with low-cost, efficient appliances might be a better option.

The search for alternative strategies to tackle the energy access problem

The worldwide disparity in primary energy (PE) consumption per capita is now a well-recognized fact (see Figure 16.4). Consumption in Africa and Asia excluding China is just 40% of the global average, estimated at 1.65 toe per person in 2002. China and Latin America also have low per capita energy consumption, between 60% and 65% of the world average.

While consumers in Africa meet their energy needs by consuming 111 kilograms of oil equivalent (kgoe) of coal, 140 kgoe of oil, 70 kgoe of gas and 314 kgoe of traditional energies, their counterparts in North America consume 1.4 toe of coal, 2.6 toe of oil, 1.5 toe of gas and 0.9 toe of other energies (including traditional energies). Thus coal consumption per person in North America is 12 times higher, while oil and gas consumption is about 20 times higher than it is in Africa.

Despite past growth and development, a large part of the world population does not have access to clean energy. The most commonly cited figures indicate that there are about 2 billion people without adequate access to clean cooking energy and about 1.7 billion without access to electricity.[9] According to the IEA in 2002,[10] half of those lacking access to electricity live in South Asia, 30% in sub-Saharan Africa and 15% in East Asia. About 85% in rural areas. Providing access to clean, affordable energy is a major challenge.

Sustainable, long-term solutions can rely neither on a subsidized supply of clean energies nor on piecemeal solutions that address only a part of the problem. What is required in the long term is an adequate supply of monetary resources to households so that they can sustain a lifestyle that relies on clean energies and other monetized inputs. Thus the energy access issue is linked to the problem of ensuring economic development, which in turn calls for an integrated approach to various development efforts at a decentralized level[11] as opposed to treating issues of electrification or energy supply in isolation. Given the diversity in terms

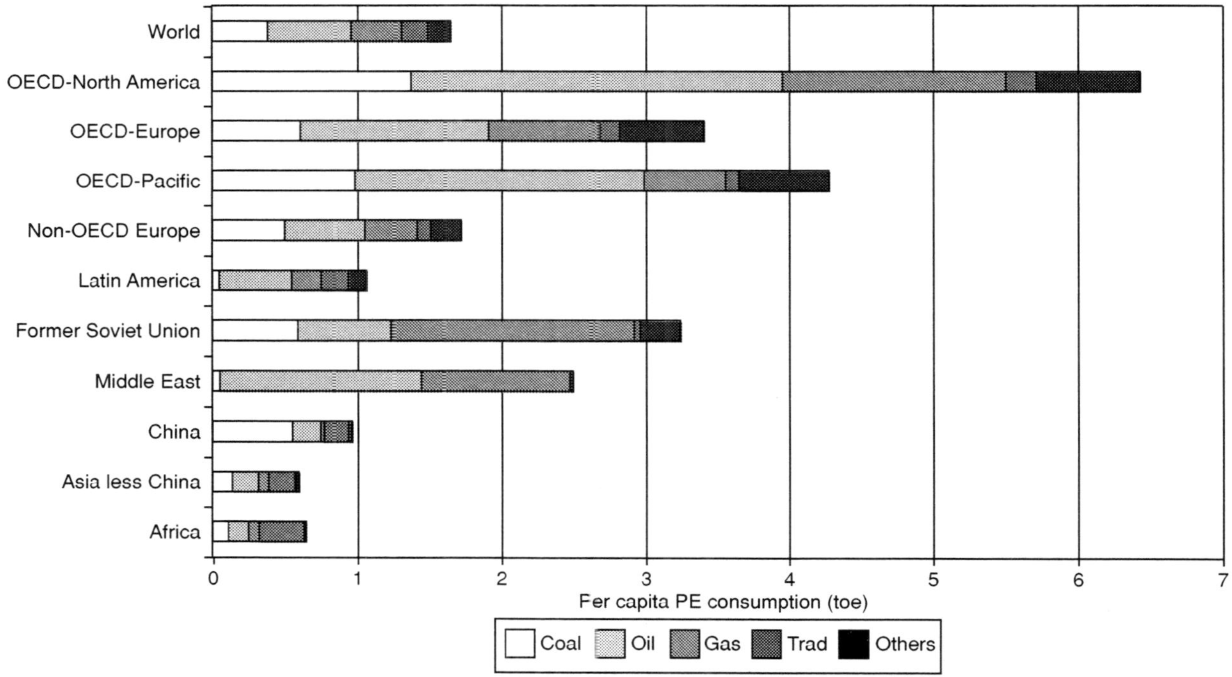

Source: IEA Key World Energy Statistics, 2004

Figure 16.4 Regional Distribution of Per Capita Primary Energy Consumption in 2002

of energy use, resource availability and other conditions, appropriate local, rather than universal or global, solutions have to be found. The policy objective should be to promote innovative solutions rather than prescribing templates. Thus each decentralized unit will have to search for its own solutions:

1. Focusing on the creation of opportunities for higher income generation in monetary terms, as opposed to in kind. Unless money flow to the poor increases, commercial energies stand little chance of competing with traditional energies.
2. Developing local energy markets, taking into account the specifics of local energy resources, needs, capacities, strengths and constraints and adopting appropriate supply mechanisms and organizational structures to cater to local needs.
3. Selective and judicious use of market interventions to make energy supply affordable but financially viable. Unless the supply is financially viable, it cannot be sustained.
4. Ensuring local community participation in the decision-making and policy-implementation process.

Such a bottom-up policy is inherently multi-dimensional and complex. It comes as a contrast to existing policies, which are top-down in nature and essentially imposed on the population. Implementation of such a policy would require the development of a common framework that could be adapted to each situation, the creation of an organizational structure to carry out the policy, building organizational capacity, adequate funding arrangements, and above all a complete review and perhaps an overhaul of the mode of functioning of the government, existing organizations and the economic activities to facilitate this decentralized approach.

Environmental concern of energy use

As different activities in the energy system (production, conversion and utilization) each have their own environmental impact, the future growth of energy demand and continued reliance on fossil fuels raise environmental concerns. As the IEA suggests,[12] 90% of energy demand is likely to be met through fossil fuels as late as 2030. The consequent environmental damage could have significant implications for humankind as a whole.

Combustion of fuels (whether in power generating stations, in industry or in residential or commercial activities) and the use of energy in transportation are the major sources of urban air pollution. Particulate matters, nitrogen oxides (NO_x), sulfur oxides (SO_x), ozone formed at ground level through the interaction of NO_x and volatile organic compounds, carbon monoxide and lead are the major pollutants that are normally monitored for health purposes.

Environmental concern has a social dimension as the poor, women and children are adversely affected due through the lack of access to clean energy discussed earlier. The use of biomass and coal in inefficient stoves and indoors without adequate ventilation at times when most of the young and vulnerable members of a community are present brings with it significant health hazards, and significant social costs, which have only recently received attention. As historically it has been found that people tend to move up the ladder of fuel quality as their income increases, once again the issue of local level environmental damage control becomes inextricable from the problem of ensuring equitable distribution of the benefits of economic development. Sector-specific intervention through the promotion of subsidized fuels or technologies leaves a heavy revenue burden on the government or the supplier.

At the same time, the urban areas of developing countries are expected to face major environmental problems. About 70% of the world's urban population lives in the developing countries of Asia, Africa or Latin America[13] and most of these urban areas suffer from poor air quality. The last hundred years have seen a dramatic change in the location of urban centers. In 1900, nine of the world's ten largest cities were in North America and Europe, whereas now only three are in the developed world.[14] Twelve of the 15 cities with the highest levels of particulate matter and six of the 15 with the highest levels of SO_x are in Asia.[15] The use of traditional energies, reliance on coal and other fossil fuel energies, and inefficient combustion techniques have degraded environmental quality. With a high level of urbanization and more energy consumption, the situation is bound to deteriorate unless active protection measures are taken through technological, economic, legal and behavioral changes.

The climate change issue which has received a disproportionate amount of attention in recent times has proved to be a challenging task given its global nature, long-term time frame (involving inter-generational issues), concern for the public good, north–south divergence of opinion and even the dispute about the scientific veracity of the problem. Consequently, the debate over whether or not to slow global warming remains inconclusive and passionate, with each party sticking to its own position. Despite some progress in the 1990s, no long-term solution has yet emerged and a direct clash of European and American approaches to decide the winner looks quite likely. Whether to rely on technological advances or to adopt economic measures to solve the problem is at the heart of this conflict.

Renewable energies: for the poor or the rich?

Renewable energy sources attracted the attention of energy policy analysts and researchers just after the oil price shocks of the 1970s. The availability of easy petrodollars, global anxiety about climate change and sustainable development, and concerns about fossil-fuel dependence and energy scarcity provided the necessary impetus to explore the potential of renewable energies. Now renewable energies occupy an important place in any strategy for sustainable development in general and sustainable energy development in particular. Yet, despite generous funding, research efforts, and policy interventions, renewable energies do not seem to enjoy much consumer patronage.

Although most renewable energies are abundantly available and have been from time immemorial, and although the poor economies are major direct users of such energy, especially for drying purposes, the more sophisticated the conversion process becomes to make it modern, the further removed it becomes from the poor. According to the IEA in 2002,[16] all such projects, promoted nationally or internationally, are supported by direct and indirect subsidies. Subsidies go against the principle of getting prices right; if a technology or energy has to depend on subsidies and favorable incentives for its financial viability, sustainable development will not result.

Renewable and decentralized energies are based on the "small is beautiful" approach as opposed to the scale and scope economy-dominated energy sector. Our very poor knowledge of how to manage small energy systems and how to organize the activities around them act as the main barriers to progress in this area. Managing such decentralized systems is very different from the traditional centralized approach. The skill set required may not be available locally and there is a need to focus on such capacity development. There also appears to be a market failure in some areas as private investors are interested in lucrative, high cost markets, not in the low marginal cost of supply (such as for solar cooking). The profitability of low marginal cost ventures being small, some alternative form of supply is required, which has not yet emerged. Traditional economics would suggest that governments should intervene in such cases but governments do not have any special skills for managing the activities of decentralized energy supply systems. Most governments work in a centralized manner and tend to impose their decisions at all levels. There may however be a case for governments providing capacity building facilities and opportunities for undertaking such activities, at least initially.

Renewable and modern energies compete with energies that involve no financial transactions. Providing modern energies almost free does not make commercial sense and does not motivate private entrepreneurs. The skills required to inaugurate an alternative form of market organization do not exist. There appears to be no niche market for renewable energies to meet the energy demand of the poor and the poor–renewable energy nexus is not a recipe for success. Additionally, the issue of path dependence is of great importance for the poor. Choosing one option automatically removes alternative options, more so for the poor, who can, of course, have severe limitations on what they can afford. Extra care should be taken in introducing strong path dependence in poor countries through deceptive and distorted incentive mechanisms that could prove costly in the long run and lead to stranded cost problems. Many renewable technologies have not yet reached maturity, and cost reductions, innovative applications and other changes cannot be ruled out. The poor, being the most vulnerable segment of society, should not be made the target of such technological experimentation, unless of course the experiments do not involve them in any financial outlay, either individually or as a nation. However, the present experience shows that the poor are the main targets and perhaps they will only benefit financially if they get such technologies free of charge.

Reform and restructuring

The history of energy industries[17] around the world is characterized by a number of distinct phases of development:

- Highly competitive markets: initially the activities started under private ownership with the objective of benefiting from the invention of a new technology, product or service. This phase was marked by cut-throat competition among players, resulting in price wars. It also saw the duplication of assets, and in some cases problems of coordination and cartelization.
- An integrated, monopolistic industry: as cut-throat competition was not conducive to taking advantage of economies of scale and scope, the next phase of development was marked by a movement towards integration and monopolization. This has taken two forms: in Europe, nationalization was the preferred option while the USA favored a regulatory approach, in which private ownership continued but profits were regulated by the government. Both

forms allowed industry consolidation and rapid progress but due to government and regulatory failures as well as developments in technologies, this form of development was questioned.
- The emergence of reforms and restructuring.

These three phases could easily be related to a number of energy industries.

Government intervention in the energy sector has been promoted either through ownership or regulatory practices or through the existence of a public good[18] argument. There are a number of alternatives to the latter:[19]

- The choice of primary energy source and technology provides greater societal benefits and costs compared to private costs. Such an argument has been used to promote nuclear energy, hydroelectric energy, and coal in various countries.
- A high share of energy industry investments in the national investment portfolio often implies that the timing and magnitude of the energy sector investments are controlled for macroeconomic policy objectives. This could offer broader social benefits.
- Private investors aim at quick recovery of their investment while public investment takes a long-term view, thereby offering social benefits.

Consequently, the energy sector is dominated by state-owned companies and the market is often regulated – either independently or otherwise.

A wave of reform, restructuring and privatization has swept the global economy since the 1980s and the energy industry could not escape. Many developing countries embraced restructuring and sector reform, often under the influence of the Bretton Woods duo and other donor countries. The general view was that performance was poor, due to a monopolistic market, state-ownership of utilities, lack of transparent regulation and political interference. All these prevented the utilities from functioning commercially, and resulted in abysmal financial health, poor service quality, inefficient system operation and poor economic growth throughout the sector. The general prescription then was to restructure through the unbundling and privatization of state-owned utilities, and the creation of an independent regulatory body.

A 2004 World Bank report states that "for much of the 1990s privatization was heralded as the elixir that would transform ailing, lethargic state enterprises into sources of creative productivity and dynamism serving the public interest." But "the privatization was oversimplified, oversold, and ultimately disappointing – delivering less than promised."[20] The same report indicates that skepticism and hostility to reform have grown and that dissatisfaction with reform and privatization is fueled by loss of jobs, price rises, high profits in the privatized companies, etc. It is now considered that in many cases reform had significant distributional impacts and that the poor and the vulnerable were affected disproportionately.

The scorecard remained unimpressive, as the reform movement did not proceed along the expected lines in many developing countries. Some provided lip service and produced precious little. The list of those following such a "wait and watch" approach is long. A few countries have completed all the milestones of the reform process as agreed with the external agency that imposed it. Yet, in terms of performance no appreciable change has been noticed and reform objectives have not been satisfied even after the completion of the project. Some other developing countries had a disastrous experience with competitive power markets. They embraced wholesale competition, provided choice of retail supplier to some consumers and restructured the entire power sector to find themselves in deep trouble in terms of long-term supply security and volatile electricity prices. The fire-fighting efforts provided wrong incentives that led to stranded costs for the sector and the country as a whole.

In the energy sector, reform and restructuring have progressed in the transition economies of Latin America, but have been less successful in the developing countries of Africa and Asia. The size of the systems has something to do with this. For example, many developing countries have very small electricity systems. The 2004 World Bank report referred to above indicates that "60 developing countries have peak system loads below 150 megawatts, another 30 between 150 and 500 megawatts, and possibly another 20 between 500 and 1000 megawatts. Even a 1000-megawatt system is small for introducing competition. Thus the benefits of competition that come from unbundling will be limited in many developing and transition economies."[21] This calls for careful balancing of the benefits and costs of unbundling on a case by case basis. This is no standard template to follow.

Experience is showing that it is difficult to ensure investment in the expansion of the network infrastructure in the unbundled industry. As the infrastructure is not yet mature in many developing countries, they need to expand the system quite rapidly to cater for growth and improvements to the existing inadequate infrastructure.

Another critical area is the regulatory capacity for ensuring proper functioning of the unbundled system. Although competition is supposed to reduce regulatory burden, the energy sector (as well as other infrastructure industries) will retain monopoly segments requiring regulatory supervision. The task becomes challenging in an unbundled environment and with low regulatory capability. It took developed countries a long time to build and develop expertise in this area. Developing countries with fewer resources will find it even more challenging.

Any reform introduces path dependence and experience shows that it is difficult to make changes to structural choices once they are introduced. Restructuring requires cooperation and participation, which may be more difficult with privatized companies than with state-owned entities.

The search for viable alternative options continues as standard reforms falter and fail to deliver in an increasing number of cases. Improved performance of the state-owned enterprises through appropriate regulation and governance would be a reasonable objective. While this area has received scant attention so far, it promises to be the subject of increased focus in the future.

Conclusion

This chapter has analyzed some major challenges facing the energy sector and suggests that the future could be dramatically different from that normally envisaged. Although depletion of reserves has generated a passionate debate about the future of oil, the chapter suggests that the lessons learned from energy transitions make it likely that an alternative will emerge before oil supplies run out. A viable substitute to oil in the transportation sector would have a serious effect on the economies of oil-rich countries, which could unleash a price war to regain market share. Both the price-war strategy and the development of a substitute for oil in transportation would bring dramatic changes to the future energy scene.

The security of energy supply is a more imminent problem than oil depletion. The traditional approach to ensuring energy supply through right prices, restructuring and the promotion of renewable energies has not succeeded. Similarly, the energy access problem cries out for new initiatives and thinking. Alternative strategies based on overall economic development, selective intervention, free trade practices and access to low cost capital and efficient technologies might be more fruitful.

Our past experience with any analysis of the future of energy suggests that the experts' consensus view has never come to pass. Nuclear was the consensus fuel of the future in the aftermath of oil shocks in the 1970s, but gas turned out to be a more important energy in reality. Renewable energies did not succeed in penetrating the energy scene while traditional sources did not retreat either. There is no reason to believe that this time around the experts won't be wrong again.

Notes

1. N. Nakicenovic, A. Grubbler and A. McDonald (eds), "Global Energy Needs: Past and Present," in *Global Energy Perspectives*, Cambridge University Press, 1998.
2. See among others, K.S. Deffeyes, *Hubbert's Peak: The Impending World Oil Shortage*, Princeton University Press, 2001.
3. North America accounts for around 5% of global oil and gas reserves while OECD Europe accounts for a miniscule 1.6% of oil and 3.4% of gas. Yet, North America produced 18% of world oil and 29% of gas in 2002. Similarly, OECD Europe produced around 9% of oil and 12% of gas in 2002.
4. IEA, *Key World Energy Statistics, 2004*, International Energy Agency, Paris, 2004. See <www.iea.org/dbtw-wpd/Textbase/nppdf/free/2004/keyworld2004.pdf>.
5. For example, in Asia without China, traditional energies play an important role and in non-OECD Europe where coal plays an important role.
6. For example, in China where coal has a 57% share, or in the Former Soviet Union where natural gas has a 52% share or in Africa where traditional energies account for 48% of the primary energy supply.
7. IEA, *World Energy Outlook*, International Energy Agency, Paris, 2002.
8. IEA, *World Energy Investment Outlook*, International Energy Agency, Paris, 2003.
9. WEA, *Energy and the Challenge of Sustainability*, World Energy Assessment 2000, UNDP, New York, 2000. Similar figures are quoted both in WEHAB, *A Framework for Action on Energy*, WEHAB Working Group, UN, New York, 2002, <www.johannesburgsummit.org/html/documents/summit_docs/wehab_papers/wehab_energy.pdf> (accessed March 25, 2004); and in DfID, *Energy for the Poor: Underpinning the Millennium Development Goals*, Department for International Development, UK, 2002.
10. IEA, *World Energy Outlook*, 2002.
11. DfID, *Energy for the Poor*; and WEC, *The Challenge of Rural Energy Poverty in Developing Countries*, World Energy Council, London, 1999.
12. IEA, *World Energy Outlook*, 2002.
13. UNEP, *Global Environment Outlook 3: Past, Present and Future Outlook*, United Nations Environment Programme, Nairobi, 2002.
14. Ibid.
15. UNEP, *Global Environment Outlook 2000*, United Nations Environment Programme, Nairobi, 2000.

16. IEA, *Financing Mechanisms for Solar Home Systems in Developing Countries: The Role of Financing in the Dissemination Process*, IEA Report PVPS T9–01:2002, Paris, 2002.
17. For a more detailed account, see IEA, *Electricity Market Reform: An IEA Handbook*, International Energy Agency, Paris, 1999. Also consult: EIA, *The Changing Structure of the Electric Power Industry: An Update*, Energy Information Administration, Washington, D.C., 1996; and D.M.G. Newbery, *Privatization, Restructuring and Regulation of Network Utilities*, MIT Press, Mass., 1999.
18. A public good is characterized by non-exclusivity and non-rivalry. Non-exclusivity implies that it is difficult to exclude others from using the good or service without withholding the good or service. Non-rivalry on the other hand means that consumption by one does not reduce its potential to be used by others.
19. M. Jaccard, "Oscillating Currents: The Changing Rationale for Government Intervention in the Electricity Industry," *Energy Policy*, Vol. 23, Issue 7, pp. 579–92.
20. World Bank, *Reforming Infrastructure: Privatization, Regulation and Competition*, World Bank, Washington, D.C., 2004.
21. Ibid.

Further Reading

While the scope of this book has been broad, it is not, and was never meant to be, universal. It would clearly be impossible to cover all topics related to the energy debate at this time, so we have focused on a few key, interrelated factors most likely to shape our common energy future. The following lists offer further reading on the following topics:

1. General background information
2. Oil-price scenarios
3. Non-conventional oil resources – tar sands, heavy oil, and oil shales
4. Renewable energy
5. Nuclear energy
6. Energy efficiency, fuel substitution and demand management
7. Public affairs / stakeholder management

1. General background information

Considine, J.I. et al. (2005) *Energy Politics*, Iss.6, Fall 2005.

Considine, J.I. and Kerr, W.A. (2002) *The Russian Oil Economy*, Edward Elgar Publishing Ltd.

Garibaldi, C.A. (2003) *Estrategia: Análisis de las Condiciones de Contorno y sus Efectos sobre Nuestra Industria*, Petrotecnia, Instituto Argentino del Petróleo y del Gas, pp.16–28.

Garibaldi, C.A. and Gerold, D.G. (2004) Sub Andean Basins and the Independents: the Future, *Oil, Gas and Energy Law Intelligence*, Vol.2, Iss.3.

IEA (2004) *Oil Crises and Climate Challenges, 30 years of Energy Use in IEA Countries*, OECD.

Lovins, A.B. et al. (2004) *Winning the Oil Endgame*, Rocky Mountain Institute.

Simmons, M.R. (2005) *Twilight in the Desert*, John Wiley & Sons.

Steger, U. et al. (2005) *Sustainable Development and Innovation in the Energy Sector*, Springer Verlag.

Tsalik, S. and Ebel, R.E. (2003) *Caspian Oil Windfalls: Who Will Benefit?* Open Society Institute.

Yergin, D. (1991) *The Prize*, Free Press.

2. Oil price scenarios

Adelman, M.A. (2002) World Oil Production and Prices 1947–2000, *Quarterly Review of Economics and Finance*, Summer 2002 Special Issue, Vol.42, Iss.2, pp.169–92.

Guidi, M.G.D., Russell, A. and Tarbert, H. (2006) The Effect of OPEC Policy Decisions on Oil and Stock Prices, *OPEC Review: Energy Economics and Related Issues*, March 2006, Vol.30, Iss.1, pp.1–18.

Jalali-Naini, A.R. and Manesh, M.K. (2006) Price Volatility, Hedging and Variable Risk Premium in the Crude Oil Market, *OPEC Review: Energy Economics and Related Issues*, June 2006, Vol.30, Iss.2, pp.55–70.
Simmons, M.R. (2006) What a Difference 20 Years Make in Crude Oil Prices, *World Oil*, Feb. 2006, Vol.227, Iss.2, pp.25–31.

3. Non-conventional oil resources

Gielen, D. and Unander, F. (2005) *Alternative Fuels: An Energy Technology Perspective*, IEA/ETO Working Paper No: ETO/2005/01, March 2005.

Tar sands

Alienating the West, (2005) *Economist*, Vol.377, Iss.8455, pp.5–7.
Another False Dawn? (2005) *Economist*, Vol.377, Iss.8451, pp.68–71.
Barker, R. (2005) Hot Bets in the Cold North, *Business Week*, Iss.3964, p.24.
Donato, K. (1998) Tapping into Tar Sands, *Northern Ontario Business*, Vol.19, Iss.2, p.27.
Parkinson, G. (2002) Combing Oil from Tar Sands, *Chemical Engineering*, Vol.109, Iss.5, p.27.
Shell Boosts Tar Sands Fleet, (2005) *E&MJ: Engineering and Mining Journal*, Vol.206, Iss.3, p.52.
Snyder, R.E. (2004) Tar Sand Recovery Chemical, *World Oil*, Vol.225, Iss.12, pp.17–17.
Stark, P.H. and Chew, K. (2005) Global Oil Resources: Issues and Implications, *Journal of Energy and Development*, Vol.30, Iss.2, pp.159–70.
Terzo, G. (2005) UBS Tapping into Alberta Tar Sands, *Investment Dealers' Digest*, Vol.71, Iss.35, p.8.
There's Oil in them Thar Sands! (2003) *Economist*, Vol.367, Iss.8330, pp.75–6.
Vardi, N. (2006) Crude Awakening (Cover Story), *Forbes*, Vol.177, Iss.6, pp.140–4.

Heavy oil

Greene, D.L., Hopson, J.L. and Li, J. (2006) Have we Run Out of Oil Yet? Oil Peaking Analysis from an Optimist's Perspective, *Energy Policy*, Vol.34, Iss.5, pp.515–31.
Heavy-Oil Cracking Process to be Demonstrated and Commercialized, (2005) *Chemical Engineering*, Vol.112, Iss.13, p.13.
Helman, C. (2004) The Big Plunge, *Forbes*, Vol.174, Iss.8, pp.90–6.
Lee, C. and Cuthill, D. (2005) Real-Time Geosteering Captures Heavy Oil Sands Value, *World Oil*, Vol.226, Iss.10, pp.33–6.
Making Light Work of Heavy Oil, (2005) *Process Engineering*, Vol.86, Iss.4, p.11.
Schramm, L.L. (2005) Oil in Saskatchewan, *Saskatchewan Business*, Vol.26, Iss.5, p.13.
Using Oil to Spread Revolution, (2005) *Economist*, Vol.376, Iss.8437, pp.33–4.

Oil shale

Big Costs for a Big Oilsands Project, (2004) *Canadian Consulting Engineer*, Vol.45, Iss.4, p. 11.

Coal City Turning to Other Industries, (2001) *Coal International*, Vol.249, Iss.5, p.190.

Costain, W.D. (1984) Energy Policy and Boom/Bust Cycles: Government Action and Instability in the Development of Oil Shale, *Policy Studies Journal*, Vol.13, Iss.2, pp.401–11.

Cummins, C. and Gold, R. (2006) Shell Buys Rights to Oil-Sands Field for $400 Million, *Wall Street Journal – Eastern Edition*, Vol.247, Iss.67, A14.

Ehinola, O.A., Shonibare, O.O. and Akanbi, O.A. (2005) Economic Evaluation, Recovery Techniques and Environmental Implications of the Oil Shale Deposit in the Abakaliki Anticlinorium, Southeastern Nigeria, *Oil Shale*, Vol.22, Iss.1, pp.5–19.

Fischer, P.A. (2005) Hopes for Shale Oil are Revived, *World Oil*, Vol.226, Iss.8, pp.69–74.

Freedenthal, C. (2005) Increased Fuel Needs Puts New Emphasis on Shale Oil, *Pipeline and Gas Journal*, Vol.232, Iss.8, p.14.

He, Y.G. (2004) Mining and Utilization of Chinese Fushun Oil Shale, *Oil Shale*, Vol.21, Iss.3, pp.259–64.

Resources and Infrastructure: Energy Provision, (2005) *Country Profile, Estonia*, pp.23–4.

Reynolds, D.B. (2005) The Economics of Oil Definitions: The Case of Canada's Oil Sands, *OPEC Review: Energy Economics and Related Issues*, Vol.29, Iss.1, pp.51–73.

Schmidt, S.J. (2003) New Directions for Shale Oil: Path to a Secure New Oil Supply Well into this Century [On the example of Australia], *Oil Shale*, Vol.20, pp.333–46.

Snyder, R.E. (2004) Oil Shale Back in the Picture, *World Oil*, Vol.225, Iss.8, p.19.

Soone, J. and Doilov, S. (2003) Sustainable Utilization of Oil Shale Resources and Comparison of Contemporary Technologies Used for Oil Shale Processing, *Oil Shale*, Vol.20, pp.311–23.

4. Renewable energy

Agbossou, K., Kolhe, M., Hamelin, J. and Bose, T.K. (2004) Performance of a Stand-Alone Renewable Energy System Based on Energy Storage as Hydrogen, *IEEE Transactions on Energy Conversion*, Vol.19, Iss.3, pp.633–40.

Barr, S., Gilg, A.W. and Ford, N. (2005) The Household Energy Gap: Examining the Divide between Habitual- and Purchase-Related Conservation Behaviours, *Energy Policy*, Vol.33, Iss.11, pp.1425–44.

Barton, J.P. and Infield, D.G. (2004) Energy Storage and its Use with Intermittent Renewable Energy, *IEEE Transactions on Energy Conversion*, Vol.19, Iss.2, pp.441–8.

Berry, D. (2005) Renewable Energy as a Natural Gas Price Hedge: The Case of Wind, *Energy Policy*, Vol.33, Iss.6, pp.799–807.

Bhattacharyya, S.C. (2006) Renewable Energies and the Poor: Niche Or Nexus? *Energy Policy*, Vol.34, Iss.6, pp.659–63.

Blok, K. (2006) Renewable Energy Policies in the European Union, *Energy Policy*, Vol.34, Iss.3, pp.251–5.

Boudri, J.C., Hordijk, L., Kroeze, C., Amann, M., Cofala, J. and Bertok, I. et al. (2002) The Potential Contribution of Renewable Energy in Air Pollution Abatement on China and India, *Energy Policy*, Vol.30, Iss.5, p.409.

Britain Trails Europe in Renewables, (2002) *Utility Week*, Vol.18, Iss.11, p.13.

Byrne, J., Kurdgelashvili, L., Poponi, D. and Barnett, A. (2004) The Potential of Solar Electric Power for Meeting Future US Energy Needs: A Comparison of Projections of Solar Electric Energy Generation and Arctic National Wildlife Refuge Oil Production, *Energy Policy*, Vol.32, Iss.2, p.289.

Cairns, R.D. (2004) Principles of Green Accounting for Renewable and Nonrenewable Energy Resources, *Energy Policy*, Vol.32, Iss.2, p.261.

China Looks to Wind Power in a Big Way, (2005) *Business and the Environment with ISO 14000 Updates*, Vol.16, Iss.12, pp.15–16.

DeCarolis, J.F. and Keith, D.W. (2006) The Economics of Large-Scale Wind Power in a Carbon Constrained World, *Energy Policy*, Vol.34, Iss.4, pp.395–410.

Dinica, V. (2006) Support Systems for the Diffusion of Renewable Energy technologies – an Investor Perspective, *Energy Policy*, Vol.34, Iss.4, pp.461–80.

Enzensberger, N., Wietschel, M. and Rentz, O. (2002) Policy Instruments Fostering Wind Energy projects – a Multi-Perspective Evaluation Approach, *Energy Policy*, Vol.30, Iss.9, p.793.

EPA Names Top Renewable Energy Users, (2006) *Professional Safety*, Vol.51, Iss.4, p.17.

Faaij, A.P.C. (2006) Bio-Energy in Europe: Changing Technology Choices, *Energy Policy*, Vol.34, Iss.3, pp.322–42.

Frey, G.W. and Linke, D.M. (2002) Hydropower as a Renewable and Sustainable Energy Resource Meeting Global Energy Challenges in a Reasonable Way, *Energy Policy*, Vol.30, Iss.14, p.1261.

Goldemberg, J. and Teixeira Coelho, S. (2004) Renewable energy – traditional Biomass vs. Modern Biomass, *Energy Policy*, Vol.32, Iss.6, p.711.

Hammons, T.J. (2001) Mitigating Climate Change with Renewable and High-Efficiency Generation, *Electric Power Components and Systems*, Vol.29, Iss.9, pp.849–65.

Higgins, A. and Koucky, S. (2003) Renewable Energy Plans Fall Short, *Machine Design*, Vol.75, Iss.4, p.58.

Huber, C. (2004) The Solar Economy: Renewable Energy for a Sustainable Global Future: Hermann Scheer; Earthscan Publications Ltd., London, September 2002, *Energy Policy*, Vol.32, Iss.3, p.437.

Isoard, S. and Soria, A. (2001) Technical Change Dynamics: Evidence from the Emerging Renewable Energy Technologies, *Energy Economics*, Vol.23, Iss.6, p.619.

Karekezi, S. (2002) Renewables in Africa – Meeting the Energy Needs of the Poor, *Energy Policy*, Vol.30, Iss.11/12, p.1059.

Kennedy, S. (2005) Wind Power Planning: Assessing Long-Term Costs and Benefits, *Energy Policy*, Vol.33, Iss.13, pp.1661–75.

Klass, D.L. (2003) A Critical Assessment of Renewable Energy Usage in the USA, *Energy Policy*, Vol.31, Iss.4, p.353.

Komor, P. and Bazilian, M. (2005) Renewable Energy Policy Goals, Programs, and Technologies, *Energy Policy*, Vol.33, Iss.14, pp.1873–81.

Lewis, D. (2004) Blooming Awful? *Power Economics*, Vol.8, Iss.4, pp.13–14.

Lugar, R.G. (2006) Thinking Outside the Barrel, *Business Week*, Iss.3977, p.124.

Mason, C.F. and Polasky, S. (2005) What Motivates Membership in Non-Renewable Resource Cartels? The Case of OPEC, *Resource and Energy Economics*, Vol.27, Iss.4, pp.321–42.
Milligan, M.R. (2002) Analytic Assessments of the Economics, Reliability, and Operating Impacts of Wind Power Plants, *Energy Studies Review*, Vol.11, Iss.1, pp.1–22.
Mitchell, C. and Connor, P. (2004) Renewable Energy Policy in the UK 1990–2003, *Energy Policy*, Vol.32, Iss.17, pp.1935–47.
Owen, A.D. (2006) Renewable Energy: Externality Costs as Market Barriers, *Energy Policy*, Vol.34, Iss.5, pp.632–42.
Owen, A.D. (2004) Environmental Externalities, Market Distortions and the Economics of Renewable Energy Technologies, *Energy Journal*, Vol.25, Iss.3, pp.127–56.
Palmer, K. and Burtraw, D. (2005) Cost-Effectiveness of Renewable Electricity Policies, *Energy Economics*, Vol.27, Iss.6, pp.873–94.
Procházková, E. (2005) Intelligent Energy for Europe – A Strategy for Permanently Sustainable Development, *Czech Business and Trade*, Iss.3/4, p.32.
Reiche, D. and Bechberger, M. (2004) Policy Differences in the Promotion of Renewable Energies in the EU Member States, *Energy Policy*, Vol.32, Iss.7, p.843.
Renewable Energy Cost Developments for Electricity Generation, (2004) *Energy Prices and Taxes*, Vol.2004, Iss.4, p.1.
Shihab-Eldin, A. (2002) New Energy Technologies: Trends in the Development of Clean and Efficient Energy Technologies, *OPEC Review: Energy Economics and Related Issues*, Vol.26, Iss.4, pp.261–307.
Siemons, R.V. (2002) How European Waste Will Contribute to Renewable Energy, *Energy Policy*, Vol.30, Iss.6, p.471.
Venetsanos, K., Angelopoulou, P. and Tsoutsos, T. (2002) Renewable Energy Sources Project Appraisal Under Uncertainty: The Case of Wind Energy Exploitation within a Changing Energy Market Environment, *Energy Policy*, Vol.30, Iss.4, p.293.
Waller-Hunter, J. (2004) "The Contribution of Renewable Energies in Meeting the Climate Challenge." Keynote Address at the International Conference for Renewable Energies. Bonn, Germany, June 1–4, 2004.
Winkler, H. (2005) Renewable Energy Policy in South Africa: Policy Options for Renewable Electricity, *Energy Policy*, Vol.33, Iss.1, pp.27–38.

5. Nuclear energy

A Landmark Nuclear Deal Agreed, (2006) *Business India Intelligence*, Vol.13, Iss.5, pp.3–4.
Allergic Reactions, (2006) *Economist*, Vol.378, Iss.8471, p.37.
Ball, J. (2006) With a Big Nuclear Push, France Transforms its Energy Equation. (Cover Story), *Wall Street Journal – Eastern Edition*, Vol.247, Iss.72, A1–A18.
Biedscheid, J.A. and Devarakonda, M. (2005) How Far are we from Public Acceptance? *Nuclear Plant Journal*, Vol.23, Iss.3, pp.36–8.
Bisconti, A.S. (2003) Why Public Support for Nuclear Energy has Increased, *Natural Gas (Wiley)*, Vol.19, Iss.10, pp.15–21.

British Dreams, (2005) *Economist*, Vol.375, Iss.8427, p.69.

Call for More Nuclear Plants, (2005) *Power Economics*, Vol.9, Iss.4, p.3.

Carey, J. (2004) Grabbing a Piece of the Nuclear Action, *Business Week*, Iss.3914, p.108.

Euro-Nuke Build Looking Inevitable, (2004) *Modern Power Systems*, Vol.24, Iss.7, p.4.

Fact and Fission, (2003) *Economist*, Vol.368, Iss.8333, pp.11–12.

Feretic, D. and Tomsic, Z. (2005) Probabilistic Analysis of Electrical Energy Costs Comparing: Production Costs for Gas, Coal and Nuclear Power Plants, *Energy Policy*, Vol.33, Iss.1, pp.5–13.

Fesmire, B. (2005) Is the Nuclear Renaissance Already Here? *Electric Light and Power*, Vol.83, Iss.5, p.47.

Ham, A. (2003) A Headache for Nuclear, *Power Economics*, Vol.7, Iss.5, p.12.

Hess, G. (2006) Bush Presses for Nuclear Recycling, *Chemical and Engineering News*, Vol.84, Iss.13, pp.34–6.

Industry Comment: UK Energy Review: Let Battle Commence, (2006) *MarketWatch: Energy*, Vol.5, Iss.3, p.5.

Krosnar, K. (2006) Nuclear "Key" to Carbon Cuts, *Utility Week*, Vol.25, Iss.5, p.11.

MacKerron, G. (2004) Nuclear Power and the Characteristics of "Ordinariness" – the Case of UK Energy Policy, *Energy Policy*, Vol.32, Iss.17, pp.1957–65.

Milne, R. (2006) Green Adviser Rejects Nuclear as Solution, *Utility Week*, Vol.25, Iss.3, p.7.

Milne, R. (2005) Nuclear Cart Put before Horse, *Utility Week*, Vol.24, Iss.17, p.8.

Nuclear Battle, (2006) *Pollution Engineering*, Vol.38, Iss.3, p.10.

Oster, S. (2006) Moscow Courts Beijing Over Nuclear Power, *Wall Street Journal – Eastern Edition*, Vol.247, Iss.68, A6.

Out of Chernobyl's Shadow, (2004) *Economist*, Vol.371, Iss.8374, pp.57–8.

Percebois, J. (2003) The Peaceful Uses of Nuclear Energy: Technologies of the Front and Back-Ends of the Fuel Cycle, *Energy Policy*, Vol.31, Iss.2, p.101.

Rao, A.R. (2006) Harnessing Nuclear Power, *Economist*, Vol.378, Iss.8470, p.18.

Robinson, S. (2006) UK Research Finds Nuclear is Not Viable, *ICIS Chemical Business*, Vol.1, Iss.10, p. 10.

Russia Needs to Build More Plants, Says Nuclear Chief, (2006) *Professional Engineering*, Vol.19, Iss.3, p.13.

Tokimatsu, K., Fujino, J., Konishi, S., Ogawa, Y. and Yamaji, K. (2003) Role of Nuclear Fusion in Future Energy Systems and the Environment Under Future Uncertainties, *Energy Policy*, Vol.31, Iss.8, p.775.

UK Energy: Nuclear Option Key to Supply Diversity, (2004) *MarketWatch: Energy*, Vol.3, Iss.11, p.8.

Vakarelska, R. and Osborne, A. (2005) Germany to Stick with Nuclear Phase-Out, *Utility Week*, Vol.24, Iss.7, p.11.

Varchaver, N. (2005) Nuclear Spring? (Cover Story), *Fortune*, Vol.151, Iss.1, p.60.

Viklund, M. (2004) Energy Policy Options – from the Perspective of Public Attitudes and Risk Perceptions, *Energy Policy*, Vol.32, Iss.10, p.1159.

6. Energy efficiency, fuel substitution and demand management

Demand Management should Alleviate Gas Dependency, (2004) *MarketWatch: Energy*, Vol.3, Iss.10, pp.12–13.

Fischer, C. (2005) On the Importance of the Supply Side in Demand-Side Management, *Energy Economics*, January, Vol.27, Iss.1, pp.165–80.

Hennicke, P. and Fischedick, M. (2006) Towards Sustainable Energy Systems: The Related Role of Hydrogen, *Energy Policy*, Vol.34, Iss.11, pp.1260–70.

Ko, J. and Dahl, C. (2001) Interfuel Substitution in US Electricity Generation, *Applied Economics*, Vol.33, Iss.14, pp.1833–43.

London Energy and Dalkia UK Try to Persuade Major Energy Buyers to Go "all-in" (2004) *MarketWatch: Global Round-up*, Vol.3, Iss.11, pp.160–1.

Ramesohl S. and Merten, F. (2006) Energy System Aspects of Hydrogen as an Alternative Fuel in Transport, *Energy Policy*, July 2006, Vol.34, Iss.11, pp.1251–9.

Sorrell, E. et al. (2004) *The Economics of Energy Efficiency: Barriers to Cost-Effective Investment*, Edward Elgar.

Sutherland, T.F. (2005) Real-Time Demand Management is Now a Reality, *Energy and Power Management*, Vol.30, Iss.7, p.8.

Utility Billing: The Smart Answer? (2005) *MarketWatch: Energy*, Vol.4, Iss.7, pp.6–7.

Zhang Shu Yu (2004) The Proposed EU Energy Security Package vis-à-vis EU Law, *European Environmental Law Review*, Vol.13, Iss.6, pp.170–6.

7. Public affairs/stakeholder management

International Public Relations Association: <www.ipra.org/>

Oil, Gas and Energy Law Intelligence: <www.gasandoil.com/ogel/>

Public Affairs Council: <www.pac.org/>

Index

Compiled by Sue Carlton